Advances in Anatomy Embryology and Cell Biology

Vol. 106

Margit Pavelka

Functional Morphology of the Golgi Apparatus

With 25 Figures

Springer-Verlag
Berlin Heidelberg NewYork
London Paris Tokyo

Dr. med. univ. Margit Pavelka

Institut für Mikromorphologie
und Elektronenmikroskopie
Universität Wien
Schwarzspanierstraße 17, 1090 Wien, Austria

ISBN 3-540-18062-1 Springer-Verlag Berlin Heidelberg New York
ISBN 0-387-18062-1 Springer-Verlag New York Berlin Heidelberg

Library of Congress Cataloging-in-Publication Data
Pavelka, Margit, 1945–. Functional morphology of the Golgi apparatus. (Advances in anatomy, embryology, and cell biology; v. 106) Bibliography: p. Includes index.
1. Golgi apparatus. I. Title. II. Series.
QL801.E67 vol. 106 574.4 s 87-16504 [QH603.G6] [574.8′734]

Typesetting, printing and binding: Universitätsdruckerei H. Stürtz AG, Würzburg
2121/3140-543210

To Ernst and Michaela

Acknowledgments

The author would like to express her deep gratitude to all members of the Institute of Micromorphology and Electron Microscopy of the University of Vienna for their thoughtfulness and generous help during the time this paper was prepared. The excellent technical assistance of Mrs. Jutta Selbmann, Mrs. Elfriede Scherzer, Mrs. Gerlinde Hartl, Mr. Helmut Oslansky, and Mr. Richard Reichhart is acknowledged with thanks. The author is also most grateful to Professor Dr. Leopold Stockinger for his generous help, critical reading of the manuscript, and continuous encouragement, and also to Professor Dr. Peter Böck for fruitful discussions and continuous stimulation. Particular thanks are due to the author's co-worker Dr. Adolf Ellinger for many hours of stimulating discussions, for the numerous suggestions he made for improvement of the manuscript, and for his generous help with its final preparation.

This work was supported by the Hochschuljubiläumsstiftung der Stadt Wien and by the Vermächtnis Josefine Hirtl.

Contents

1 Introduction

The complex apparatus of stacked cisternae, tubules, and vesicles, "il apparato reticulare interno" (Fig. 1), first described by Camillo Golgi in 1898 (Golgi 1898), then neglected during many years, has been rediscovered and in the last decades been proved a central crossroad in intracellular traffic. Elements of the Golgi apparatus are important stations in the routes of newly synthesized molecules, including secretory molecules, membrane constituents, and lysosomal enzymes, as well as internalized molecules (Fig. 2; for recent reviews, see Bennett 1984; Dunphy and Rothman 1985; Farquhar 1985; Farquhar and Palade 1981; Goldfischer 1982; Hand and Oliver 1981; Morré and Ovtracht 1977; Palade 1983; Rothman 1985; Slot and Geuze 1983; Tartakoff 1980, 1983a; Völkl 1980; Whaley and Dauwalder 1979).

Flat cisternae (saccules), interconnected by tubular-reticular elements, are arranged in parallel to form a system of stacks, the characteristic "Golgi stacks" (dictyosomes; Fig. 3), which represent morphologic subunits of the complex organelle. The stacked cisternae are the sites where multiple posttranslational modifications occur at newly synthesized molecules, such as insertion of terminal

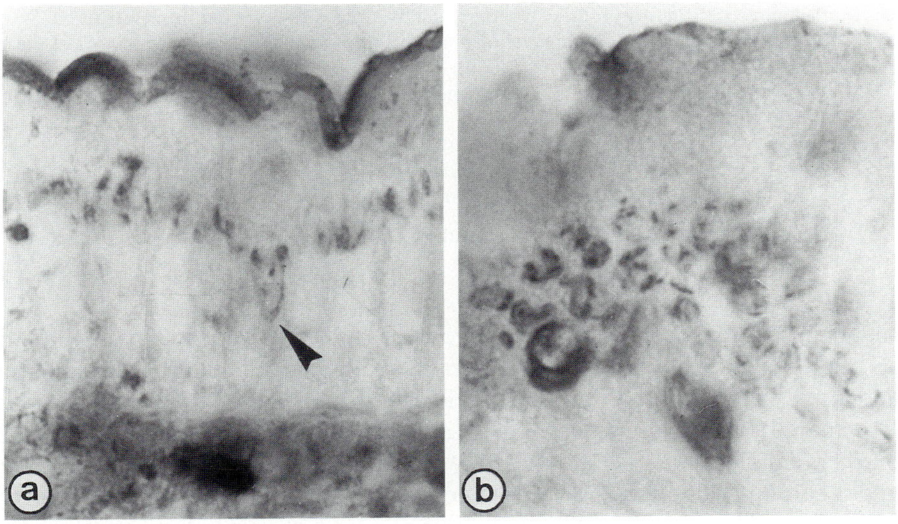

Fig. 1 a, b. Epithelium of the rat small intestine. Affinity cytochemical labeling of the Golgi apparatus. HPA-HRP conjugates. In **a**, the epithelium is sectioned perpendicularly to the surface showing the characteristic supranuclear position of the Golgi apparatus in the absorptive cells and in a goblet cell (➤); in the oblique section in **b**, the reticular, ring- or gobletlike arrangement of the Golgi elements is apparent. **a** ×1600; **b** ×1800

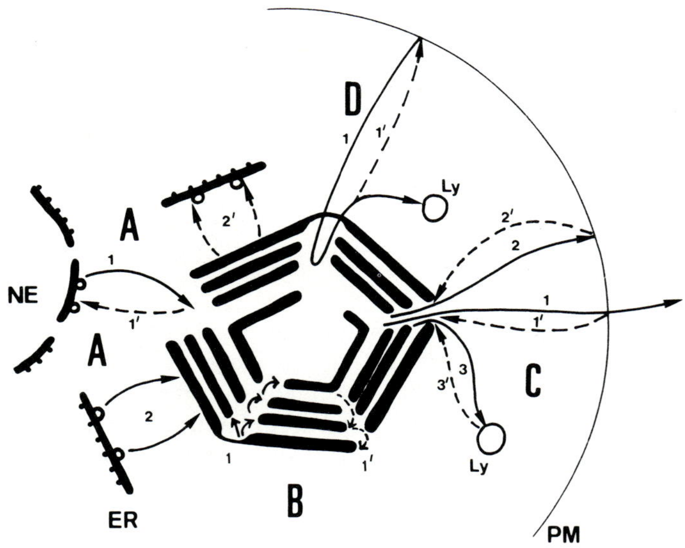

Fig. 2. Cellular transport routes involving elements of the Golgi complex.

A *1*, Nuclear envelope-to-Golgi route of newly synthesized molecules.
2, Endoplasmic reticulum-to-Golgi route of newly synthesized molecules.
B, *1*, Golgi intercompartmental route; traffic of newly synthesized molecules from one Golgi subcompartment to others.
C, Golgi-to-final destination route of newly synthesized molecules.
1, Golgi-to-cell surface route of secretory molecules.
2, Golgi-to-plasma membrane route of plasma membrane molecules.
3, Golgi-to-lysosome route of lysosomal enzymes.
D, *1*, Cell surface-to-Golgi route of internalized molecules.
Recycling routes.
D, *1′*, Golgi-to-plasma membrane recycling route of plasma membrane receptors.
C, *1′*, Plasma membrane-to-Golgi recycling route of molecules involved in secretory pathways.
2′, Plasma membrane-to-Golgi recycling routes of molecules involved in transport of plasma membrane constituents.
3′, Lysosome (or prelysosomal compartment)-to-Golgi route of receptors for lysosomal enzymes.
B, *1′*, Golgi intercompartmental recycling route.
A, *1′*, Golgi-to-nuclear envelope recycling route.
2′, Golgi-to-endoplasmic reticulum recycling route.

NE, nuclear envelope; *ER*, endoplasmic reticulum; *PM*, plasma membrane; *Ly*, lysosome

sugars into N-glycosidically linked oligosaccharides of glycoproteins and synthesis of O-glycosidically linked glycans (reviewed in e.g., Fleischer 1983; Hubbard and Ivatt 1981; Kornfeld and Kornfeld 1985; Northcote 1979; Schachter and Roseman 1980; Sturgess et al. 1978), formation of the lysosomal recognition marker (e.g., Pohlmann et al. 1982; Sly and Fischer 1982), sulfation (e.g., Fatem and Leblond 1985; Herzog 1985; Lee and Huttner 1985; Young 1973), and proteolytic cleavage (e.g., Orci et al. 1985b). Golgi-associated processes include steps in the synthesis and transport of proteoglycans (e.g., Kimura et al. 1984; Ratcliffe et al. 1985; Stow et al. 1985) and lipids (Keenan et al. 1974; Lipsky and Pagano 1985a, b; Yusuf et al. 1983, 1984). Final formation and packaging of lipoprotein particles (e.g., Banerjee and Redman 1984; Christensen et al.

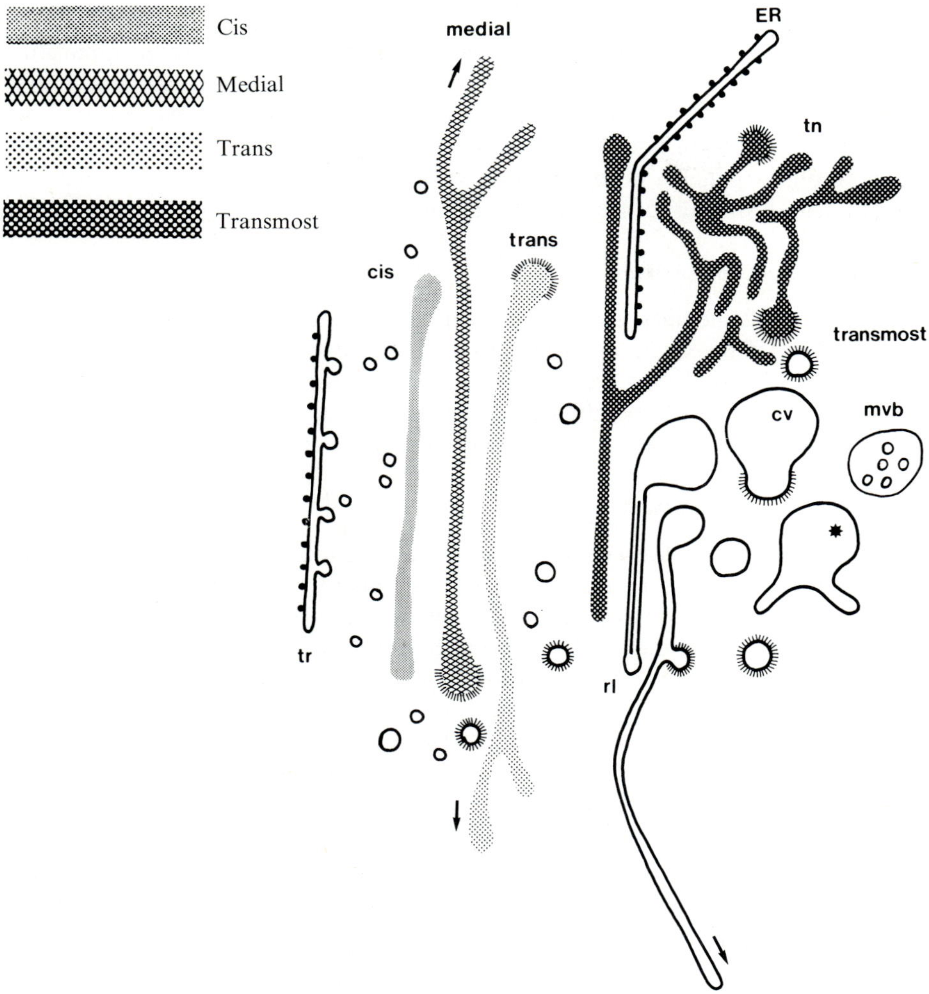

Fig. 3. Simplifying diagram of a Golgi stack.

cis, cis subsection *medial*, medial subsection *trans*, trans subsection *transmost*, transmost subsection	of the stack

ER, endoplasmic reticulum; *tr*, transitional element of the endoplasmic reticulum; *rl*, rigid lamella; *cv*, condensing vacuole; *tn*, trans Golgi tubular network; *mvb*, multivesiculated body; *, vesicle with fingerlike extensions; → interstack bridge; cytoplasmically coated membrane segment

1983; Friedman and Cardell 1976; Higgins and Hutson 1984; Howell and Palade 1982; Jersild 1966; Jones and Ockner 1971; Matsuura and Tashiro 1979; Morré and Ovtracht 1981; Sabesin and Frase 1977), condensation and forwarding of secretory materials (e.g., Bendayan 1984; Jamieson and Palade 1967a, b, 1968; Kelly 1985; Palade 1975), targeting of plasma membrane proteins (e.g., Bennett et al. 1974; Bergeron et al. 1982b; Bergmann et al. 1981; Fambrough and Devreotes 1978; Tokumitsu and Fishman 1983), as well as packaging of lysosomal enzymes and formation of primary lysosomes (e.g., Brown and Farquhar 1984a; Novikoff 1976) take place in elements of the Golgi apparatus. Furthermore, Golgi elements are recipients of internalized molecules and implicated in the recycling routes of plasma membrane constituents (for review, see Farquhar 1983, 1985; Morré et al. 1984a; Pastan and Willingham 1985). In the Golgi organelle, newly synthesized as well as internalized molecules are partitioned according to their cellular destinations.

Functional Subcompartments

A number of biochemical and morphologic studies indicate a compartmentalized organization of the Golgi apparatus (for recent reviews, see Berger 1985; Dunphy and Rothman 1985; Tartakoff 1983a); however, we do not yet know in which way functional subcompartments are arranged in the complex Golgi system and how they cooperate to build up functional units.

Morphologic Subsections

The morphologic appearance of the Golgi stacks and reaction patterns obtained by immunocytochemistry and lectin cytochemistry point to the existence of at least four subsections of the Golgi stacks (Fig. 3), viz., the cis (proximal, forming, immature), the medial (intermediate, intercalary), the trans (distal, secretory, mature), and the transmost (referred to as the GERL system by Novikoff 1964; trans-tubular network by Rambourg et al. 1979; trans Golgi reticulum by Willingham and Pastan 1984) subsections. As to whether the Golgi stack subsections that are apparent morphologically correspond to functional Golgi subcompartments is one of the crucial questions concerning Golgi organization. Certain processes, e.g., successive steps in the biosynthesis of glycans, have been associated with cisternae of certain definite subsections of the stacks, the sequence of events being oriented from the cis to the trans side (reviewed in Berger 1985; Dunphy and Rothman 1985). In contrast, other studies have revealed a widely uniform appearance of the stacked cisternae (e.g., Hedman et al. 1986; Novikoff et al. 1983b; Pavelka and Ellinger 1986b–d; Roth et al. 1986) or have indicated heterogeneous and variable arrangement of subcompartments (e.g., Oliver and Hand 1983; Pavelka and Ellinger 1986c) and/or the existence of functional subdomains within the individual cisternae (reviewed in Farquhar and Palade 1981).

In the last few years, a number of comprehensive Golgi apparatus reviews have been published (Bennett 1984; Berger 1984, 1985; Dunphy and Rothman 1985; Farquhar 1985; Farquhar and Palade 1981; Goldfischer 1982; Hand and Oliver 1981; Morré and Ovtracht 1977; Robinson and Kristen 1982; Rothman 1985; Slot and Geuze 1983; Tartakoff 1980, 1983a; Whaley 1975; Whaley and Dauwalder 1979). The present paper does not want to compete with these presentations; it is dedicated to functional morphology of the Golgi apparatus, giving prominence to morphologic and cytochemical aspects of Golgi organization. The author's studies concentrated particularly on the architecture of the Golgi apparatus as found in goblet and absorptive cells of the small intestine and in pancreatic acinar cells.

2 Material and Methods

Female albino rats (Him: OFA [SD] SPF), including mated animals, were obtained from the "Forschungsinstitut für Versuchstierzucht" of the University of Vienna. The animals were fed a standard diet (Altromin 1314, 1324) and had free access to water. A fasting period of 24 h preceded those experiments which were concentrated on intestinal and pancreatic tissue. In general, the tissues were excised under anesthesia with pentobarbital.

For studies of embryonic pancreatic tissue, at the appropriate gestational ages, i.e., daily from day 13 to day 20 of gestation, caesarean section was performed under anesthesia with pentobarbital; the pancreatic rudiments were fixed in situ (fixation procedures, see below) and excised with the aid of a preparation microscope.

Colchicine Experiments

Colchicine (Fluka AG, Basel, Switzerland) was dissolved in 0.9% NaCl immediately before use and injected intraperitoneally at a dosage of 0.5 mg/100 g of body weight; the time of administration was consistently 9 a.m. Control animals received an aliquot of 0.9% NaCl. The experiments included morphologic and enzyme-cytochemical studies of the small intestinal epithelial and pancreatic acinar cells at 10, 20, 30, and 45 min and 1, 2, 4, 5, and 6 h after administration of colchicine.

Morphology

For morphologic examinations, the tissues were fixed in 2.5% glutaraldehyde (electron microscopy grade, Merck, Darmstadt, FRG) in 0.1 M sodium cacodylate buffer, pH 7.2, for 2 h at 4° C, postfixed in 1% veronal acetate-buffered OsO_4, dehydrated in a graded series of ethanol, and embedded in Epon. Thin sections were stained in alcoholic uranyl acetate and alkaline lead citrate and examined in a Zeiss EM 9 or a Philips EM 400 electron microscope.

Cytochemistry

Prolonged Osmification. Small pieces of tissue were immersed in 2% aqueous OsO_4 and incubated for 40 h at 40° C (Friend and Murray 1965).

Enzyme Cytochemistry. The tissues were fixed in 2.5% glutaraldehyde (electron microscopy grade, Merck, Darmstadt, FRG) for 1 h at 4° C; after an overnight rinse in 0.1 M sodium cacodylate containing 10% dimethyl sulfoxide (DMSO) and 7.5% sucrose, 30- to 40- µm-thick cryosections were prepared.

Thiamine pyrophosphatase (TPPase) and inosine diphosphatase (IDPase) were demonstrated according to the method of Novikoff and Goldfischer (1961): the medium consisted of 25 mg thiamine pyrophosphate or inosine diphosphate, respectively, 7 ml bidistilled water, 10 ml 0.2 M tris-maleate buffer, pH 7.2, 5 ml 0.025 M manganese chloride, 3 ml 1% lead nitrate, and 1.25 g sucrose; the cryosections were incubated for 70 min at 37° C.

Acid phosphatase (AcPase) was demonstrated according to Barka (1964): the cryosections were incubated in a medium consisting of 10 ml 1.25% sodium-β-glycerophosphate, 10 ml 0.1 M tris-maleate buffer, pH 5.0, 10 ml bidistilled water, 20 ml 0.2% lead nitrate, and 7.5% sucrose for 30 min at 37° C.

For localization of trimetaphosphatase (TMPase), the cryosections were treated according to Doty et al. (1977): the incubation medium consisted of 18 mg sodium trimetaphosphate,

4.5 ml 0.1 M acetic acid, 50 ml bidistilled water, and 10 ml 1.5% lead acetate; the pH was adjusted to 3.9 with 0.1 M nitric acid and the volume of the medium raised to 100 ml by adding distilled water; finally, 5 g sucrose were added.

For cytochemical controls, the sections were incubated in equivalent media lacking the substrate.

Following several rinses in bidistilled water and postfixation in either 1% veronal-acetate-buffered OsO_4 or 1% osmium ferrocyanide for 1 h at 4° C, the sections were dehydrated in ethanol and embedded in Epon.

Lectin Cytochemistry. After a 1- to 2-min rinse in 0.1 M sodium cacodylate buffer, pH 7.2, fixation of the tissues was performed in a mixture of 4% paraformaldehyde and 0.5% glutaraldehyde for 30 min at 4° C; following several rinses in buffer, free aldehyde groups were blocked by treatment with 0.05 M ammonium chloride for 2 h at 20° C.

Preembedment lectin labeling. For preembedment lectin staining, a procedure was performed that was a modification of a technique originally described by Bernhard and Avrameas (1971). The tissues were rinsed in 0.1 M sodium cacodylate containing 10% DMSO for at least 24 h. Cryosections 10 μm thick were prepared, thawed in phosphate-buffered saline (PBS), and incubated in solutions in PBS of the respective lectins or lectin-horseradish peroxidase (HRP) conjugates for 4 h at 20° C: concanavalin A (ConA, Sigma Chem. Co., St. Louis, MO; 200 μg/ml) followed by treatment with HRP (grade VI, Sigma: 200 μg/ml), *Ricinus communis I* (RCA I)-HRP (E-Y Laboratories, San Mateo, CA; specificity, β-D-galactose > α-D-galactose; 50–100 μg/ml), *Helix pomatia* lectin (HPA)-HRP (E-Y Laboratories; specificity, α-N-acetyl-D-galactosamine > α-N-acetyl-D-glucosamine > α-D-galactose; 50–100 μg/ml), *Pisum sativum* lectin (PSA)-HRP, and *Lens culinaris* (LCA)-HRP (E-Y Laboratories; specificity, α-D-mannose, α-D-glucose, for high-affinity binding to glycopeptides, a fucose residue attached to the asparagine-linked N-acetyl-glucosamine being essential; Debray et al. 1981; Kornfeld et al. 1981; 50–100 μg/ml). The incubation media, in addition, contained 0.1 mg/ml saponin and 1 mg/ml bovine serum albumin.

After an overnight rinse in PBS, the peroxidase activity was visualized by means of the diaminobenzidine (DAB) reaction: the cryosections were incubated in 0.05 M tris HCl buffer, pH 7.6, containing 0.5 mg/ml DAB and 20 μl 1% H_2O_2 for 20 min at 20° C. The specimens were postfixed in 1% osmium ferrocyanide for 30 min at 4° C followed by incubation in 1% veronal acetate-buffered OsO_4 for at least 8 h at 4° C, dehydrated in ethanol, and embedded in Epon. Thin sections were examined unstained.

Postembedment lectin labeling. For post-embedment lectin staining, small pieces of tissue that were fixed and treated with ammonium chloride, as described above, were embedded in the hydrophilic resin "LR-White" (London Resin Comp., Basingstoke, UK; Ellinger and Pavelka 1985; Newman et al. 1983). Thin sections were prepared and, without using a supporting film, mounted on gold or nickel grids. For lectin labeling, the grids were immersed in or floated on a drop of the incubation medium. Colloidal gold conjugates of the lectins purchased from E-Y Laboratories (San Mateo, CA) were used; the sections were incubated in the respective lectin-gold conjugates at concentrations 40–80 μg/ml for 1 h at room temperature. After several rinses in bidistilled water, the sections were counterstained in 2% aqueous or alcoholic uranyl acetate/2% aqueous lead citrate or in 1% phosphotungstic acid.

For cytochemical controls, in the case of preembedment incubations, the lectin incubation step was omitted prior to treatment with HRP (200 μg/ml, 4 h, 20° C). For both the pre- and postembedment procedures, the specificity of the reactions was tested by adding the respective competitive sugars (0.2–0.4 M) to the incubation media: methyl-α-mannoside, ConA, PSA, LCA; D-galactose, RCA I; N-acetyl-D-galactosamine, HPA.

Endocytosis Studies

Endocytosis experiments were performed with myeloma cells (a murine cell line, P3X, and a human cell line, RPMI, both kindly provided by Prof. Heinz Ludwig, II. Medical Clinic, University of Vienna; uptake of cationized ferritin [CF], purchased from Miles Laboratories Inc.) and rat fibroblasts (W1, kindly provided by Dr. Monika Vetterlein, Institute of Tumor Biology, University of Vienna; uptake of PSA and LCA-HRP conjugates, purchased from E-Y Laboratories). CF as well as PSA- and LCA-HRP conjugates were added to the incubation media at concentrations of 50–100 μg/ml. Incubations were at 37° C in an atmosphere of

95% O_2 and 5% CO_2 for 15, 30, 60, and 120 min. Cells exposed to CF were fixed in 2% glutaraldehyde in 0.1 M sodium cacodylate, pH 7.2, for 1 h at 4° C; those exposed to PSA/LCA-HRP conjugates were fixed in a mixture of 4% paraformaldehyde and 0.5% glutaraldehyde for 30 min at 4° C. In the case of the PSA/LCA-HRP uptake studies, the peroxidase activity was demonstrated by the DAB reaction as described above. After postfixation in 1% veronal acetate-buffered OsO_4 for 1 h at 4° C, the cells were dehydrated and embedded in Epon.

3 Architecture of the Golgi Apparatus

The Golgi apparatus (Golgi complex, Golgi organelle, Golgi body) constitutes a complex intracellular system of flat cisternae (saccules), tubules, and vesicles that is preferentially located in supra- and perinuclear cytoplasmic regions (the "Golgi region," "Golgi area"; Fig. 1).

Sets of cisternae are arranged in parallel, thus building up the characteristic "Golgi stacks" (dictyosomes, e.g., Figs. 3, 4), which represent morphologic subunits of the complex apparatus; a variable number of stacks, which are interconnected, constitutes the overall Golgi organelle. The interconnections of the stacks may be constructed by cisternae which are particularly elongated and form a kind of "interstack bridge" (Fig. 6). Otherwise, neighboring stacks may be interconnected by extensive tubular-reticular elements (Fig. 8b). For some cell types, this has been impressively shown by examining several-micron thick sections and using high-voltage electron microscopy (e.g., Carasso et al. 1971; Hermo et al. 1980; Noda and Ogawa 1984; Novikoff et al. 1971; Rambourg and Clermont 1986; Rambourg et al. 1981, 1984) as well as by high-resolution scanning electron microscopic investigations (Tanaka et al. 1986). These studies have made it evident that the Golgi apparatus is a continuous organelle of reticular, ring- or gobletlike architecture.

The location in distinct cytoplasmic areas of the Golgi apparatus is closely related to its special tasks (e.g., Clermont and Tang 1985; Ede and Wilby 1981; Garcia-Porrero et al. 1981; Geuze and Kramer 1974; Herzog and Farquhar 1977; Kupfer et al. 1982; Lucocq and Montesano 1985; Nemere et al. 1985; Tassin et al. 1985). The location of the stacks of cisternae may change in response to alterations of the cell function and is decisive for the smooth functioning of several Golgi-connected processes (e.g., Rogalski et al. 1984).

3.1 Golgi Stacks

A variety of cell types exhibits a clear polar construction of the stacks of Golgi cisternae (Figs. 4, 5, 10, 11, 15–18, 22): the morphologic characteristics and

Fig. 4a, b. Stacks of Golgi cisternae of pancreatic acinar cells of the rat. At the cis side (*cis*), the cisternae are irregularly dilated, whereas they show a narrow, more regular luminal space with dense contents at the trans side (*trans*). In **a**, numerous small vesicles, presumably representing transport vesicles (→), are apparent between transitional elements of the endoplasmic reticulum (←) and cis Golgi cisternae, as well as close to the poles of the stacks. In the trans region, cytoplasmically coated circular and oval membrane profiles are abundant. In **b**, one trans saccule terminates in a balloon-like dilatation (→), presumably representing the budding-off of a secretory granule. *cv*, condensing vacuole. **a** ×51 000; **b** ×51 000 ▷

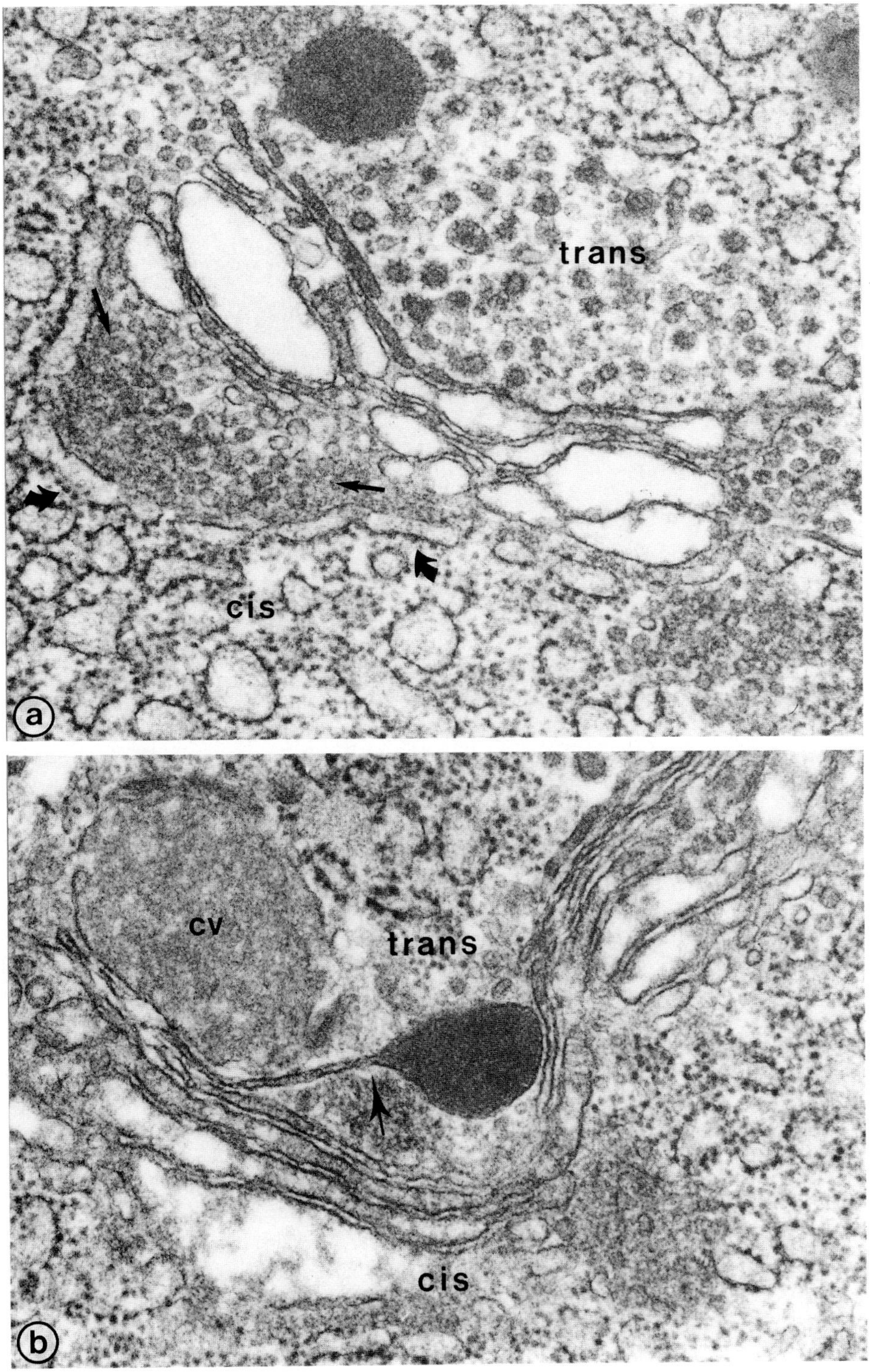

Fig. 4a, b

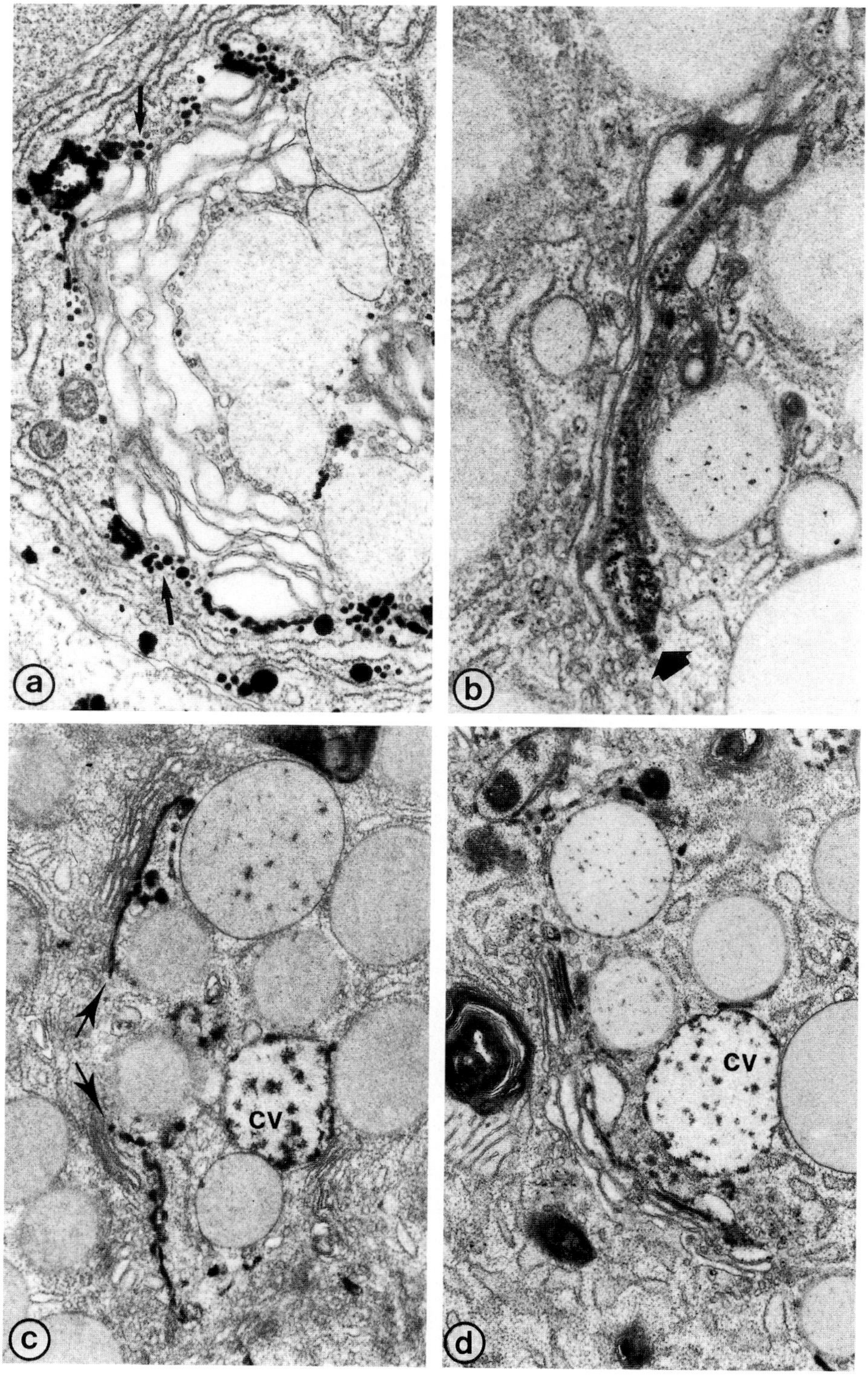

Fig. 5 a–d

cytochemical reactions of the cisternae of one side differ from those of the other side. In addition, they frequently contrast to those of the intercalary ones; this makes it possible to distinguish between two sides, viz., cis and trans, and four subsections, viz., cis, medial, trans, and transmost, of the Golgi stacks (Fig. 3). The wide, irregularly dilated saccules of the cis side stand in contrast to the narrow trans cisternae (e.g., Figs. 4, 10a, 11b, c). In addition, the trans side is characterized by special structures, such as "rigid lamellae" and tubular-reticular elements that are mostly located at some distance from the stacks (e.g., Figs. 6, 7a, 8, 9; cf. Sect. 3.1.2).

The cisternal plates exhibit a variable number of pores (e.g., Rambourg et al. 1981; Fig. 8b); at their rims, the cisternae may be continued by tubular elements that interconnect the individual stacks and, at the trans side, correspond to the special "trans tubular reticulum" (Fig. 8a, c; cf. pp. 19–21). Interconnecting elements, such as tubules and cisternae, do not only exist between the corresponding subsections of neighboring stacks, e.g., represent cis-to-cis interconnections, but are apparent also between cisternae of different levels of the stacks, e.g., connect medial cisternae of one stack with trans cisternae of a neighboring stack (e.g., Hermo et al. 1980; Rambourg et al. 1979; Tanaka et al. 1986). The cisternae, particularly those of the trans side, may terminate in vacuolar distensions or knoblike buds, which are considered as mirroring the pinching-off of secretion granules and diverse Golgi vesicles (Figs. 4, 6a, b).

Frequently, though not constantly, the classic cytochemical Golgi reactions (cf. p. 33) confirm the polar construction of the stacks: osmium deposits after prolonged osmification (Friend and Murray 1965) are preferentially cis located (Fig. 5a). TPPase, nucleoside diphosphatase (NDPase), and AcPase activities predominate at the trans side (Fig. 5b–d; Hand 1980; Novikoff and Novikoff 1977), and nicotinamide adenine dinucleotide phosphatase reactions (Smith 1980) stain medial cisternae.

In some kinds of cells, it has been shown that the thickness of the membranes of the Golgi cisternae increases and staining characteristics change from the cis to the trans side of the stacks (e.g., Grove et al. 1968; Mollenhauer et al. 1976; Morré et al. 1979; Pelttari and Helminen 1983; Stanka et al. 1981). Freeze-fracture studies (Sesso et al. 1983; Staehelin and Kiermayer 1970; Volkmann 1983) have revealed that the number of intramembrane particles differs between the cisternae of different regions of the stacks. By means of the filipin method, the lipid composition of the Golgi membranes in endocrine pancreatic cells has been shown to be heterogeneous, the cholesterol content being higher in the membranes of the trans cisternae than in those of the cis cisternae (Orci et al. 1981).

◁ **Fig. 5a–d.** Golgi stacks of pancreatic acinar cells of the rat. **a** Prolonged osmification. Osmium deposits label transport vesicles (→) and one cisterna at the cis side of this stack. ×16800. **b** TPPase. TPPase reaction products stain one trans saccule (➧). ×44000. **c** AcPase. AcPase reaction is confined to the Golgi elements located in the transmost position of these stacks: the transmost cisternae (→), vesicles of the trans Golgi side, and condensing vacuoles (*cv*). ×23000. **d** IDPase. IDPase reaction products label trans Golgi cisternae and condensing vacuoles (*cv*). ×31500

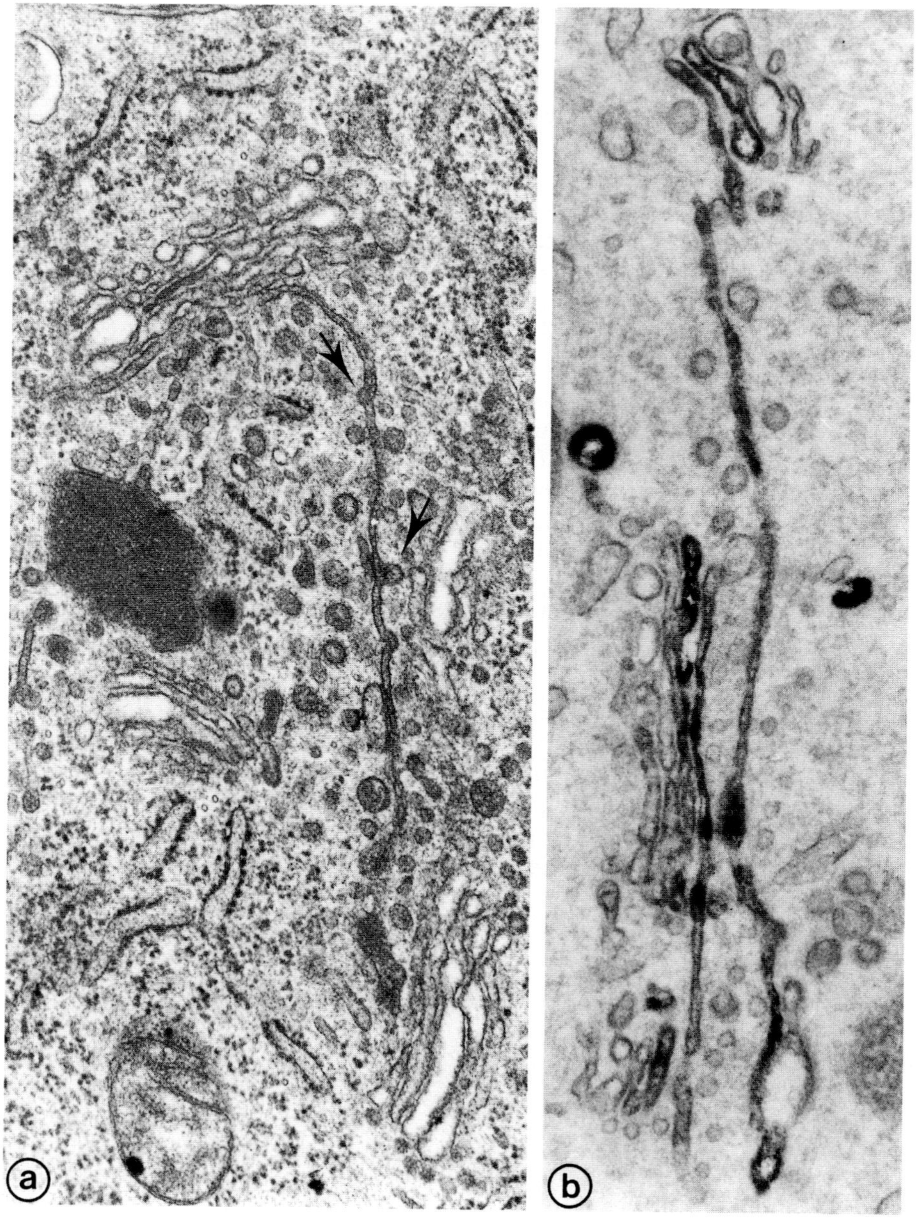

Fig. 6a, b. Golgi stacks of acinar cells of the rat embryonic pancreas in the early differentiated state (days 15/16 of gestation). **a** Morphology. **b** RCA I staining. An elongated cisterna, in part exhibiting a largely regular luminal width reminiscent of rigid lamellae, is located in the transmost position of three (in **a**) and two (in **b**) Golgi stacks and forms a kind of interstack bridge; knoblike protuberances (→) may mirror the budding-off of vesicles. RCA I binding reaction (in **b**) labels the prominent transmost cisterna as well as saccules of the medial/trans subsections of the stacks. In both figures, cytoplasmically coated and smooth vesicular structures are abundant near the trans Golgi face. **a** ×37000; **b** ×44000

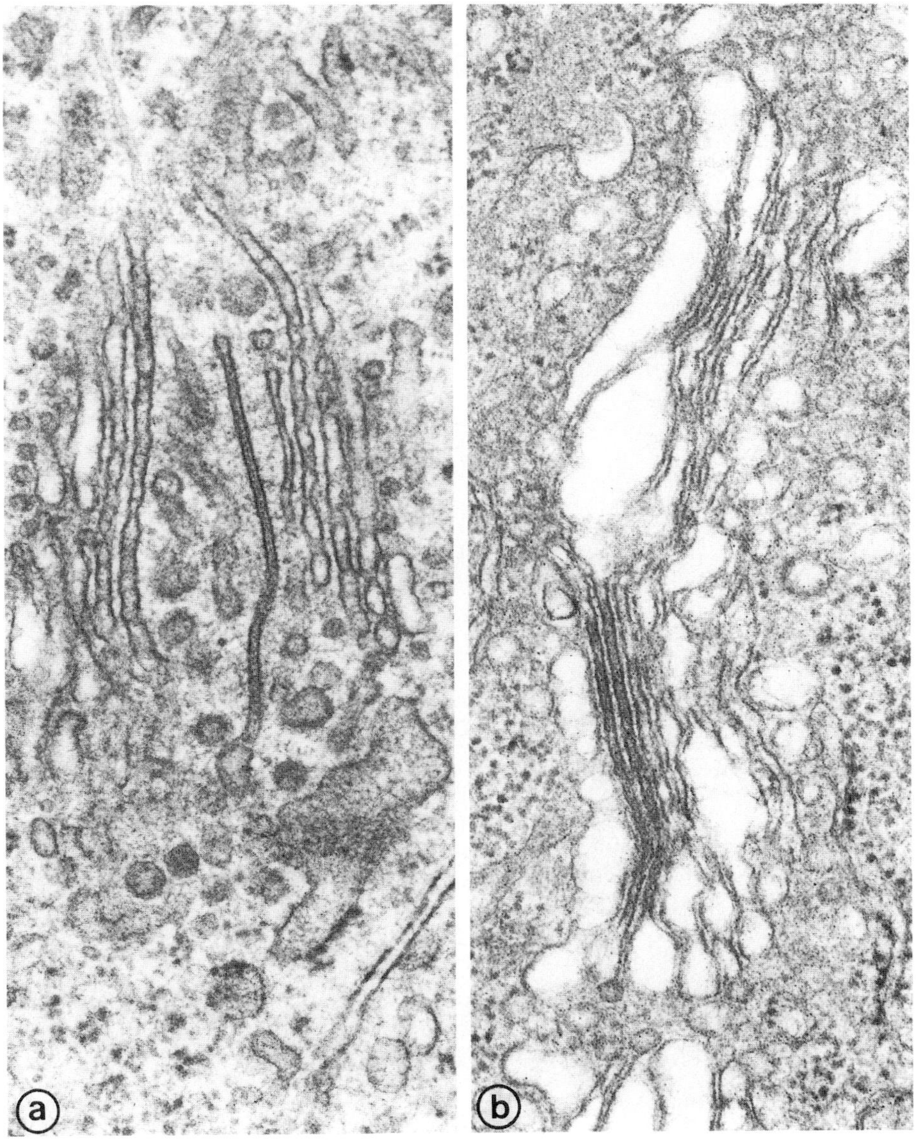

Fig. 7a, b. Rigid lamellae. **a** Golgi stack of an embryonic pancreatic acinar cell of the rat at day 16 of gestation. The transmost cisterna of this stack is located somewhat detached from the other saccules and is characterized by an extremely regular luminal space containing dense materials (midline densities). ×61 000. **b** Golgi stack of a rat small intestinal absorptive cell of the crypt-top region. Cisternae with an extremely regular luminal width and midline densities, reminscent of rigid lamellae, are apparent in the medial subsection of this stack. ×61 000

Membrane continuities between the individual cisternae of one given stack have been found in some cell types. Such continuities, being difficult to recognize in ultrathin sections, have been demonstrated rarely by transmission electron microscopy (e.g., Bracker et al. 1971). High-resolution scanning electron microscopic studies have confirmed the existence of such interconnections and have

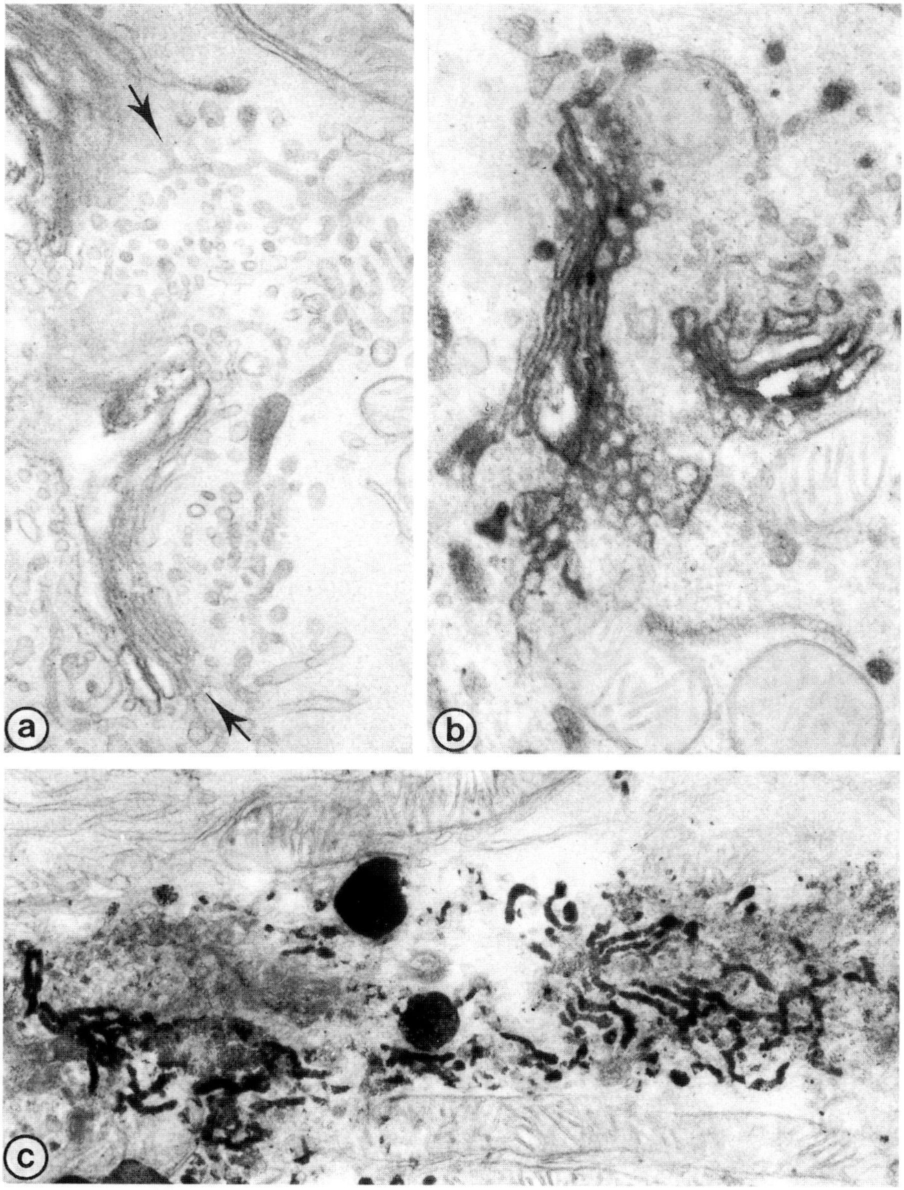

Fig. 8a–c. Epithelial cells of the crypt-top region of the rat small intestine. Tubular-reticular Golgi elements. **a** Cisternal plates continue into tubular elements (→) that constitute a tubular-reticular system at the trans side of this stack. ×34000. **b** LCA binding reaction. A cisterna with numerous pores forms a kind of interstack bridge between two neighbored Golgi stacks. ×34000. **c** Acid phosphatase. A network is formed by continuous tubular elements being labeled by acid phosphatase reaction products. ×27000

Fig. 9a–c. Fibroblasts. Internalization of HRP-labeled PSA, 60 min. Reaction products, localizing internalized PSA-HRP conjugates, are concentrated in elements residing in the transmost position of these Golgi stacks: extended tubular-reticular elements (→), multivesiculated bodies (➘), and polymorphous vacuoles with fingerlike extensions (*). Some of the reactive tubules and vesicles are located near the cis face of the stacks (▷). **a** ×42500; **b** ×33500; **c** ×37000 ▷

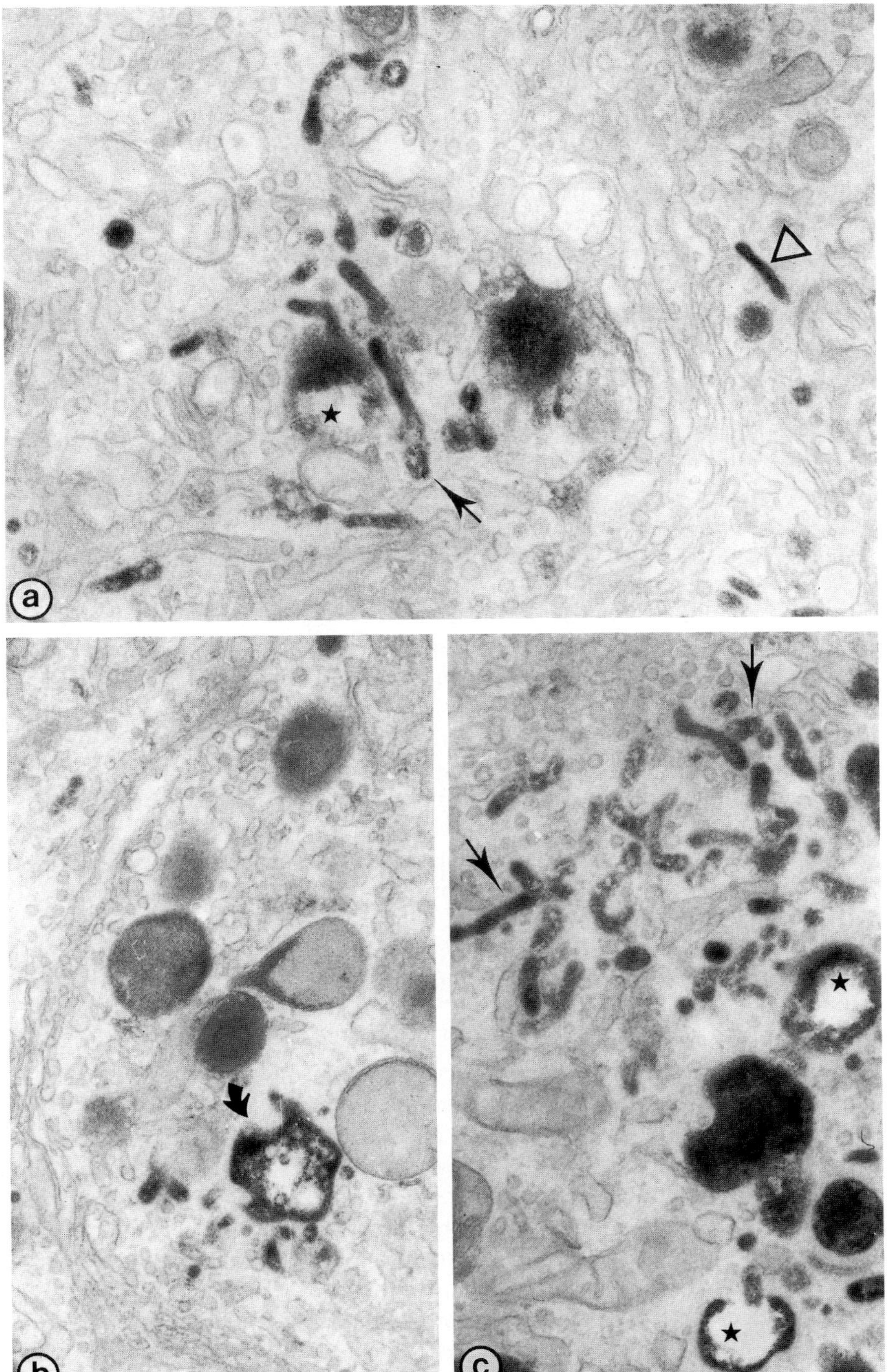

Fig. 9 a–c

indicated that Golgi stacks may be constructed by one single, helically wound cisterna (Tanaka et al. 1986).

3.1.1 Transitional Elements – Transport Vesicles

At the cis face, as well as close to the poles, transitional elements of the endoplasmic reticulum, with polyribosomes attached to one surface only, frequently adjoin the stacks (Fig. 4a, b). They may show a budding-off of small vesicles measuring 50–80 nm in diameter (transport vesicles, primary vesicles, transition vesicles). In some cell types, such as in the embryonic cells of the grasshopper (Kessel 1971), transport vesicles bud off from the nuclear envelope. Small transport vesicles are particularly abundant near the cis cisternae (Figs. 4a, 18b), close to the poles of cisternae, and in the zones in between the stacks (Figs. 19b, 20, 23; "passageways," Novikoff et al. 1977). Furthermore, small vesicles have been found associated with pores and larger perforations of the cis saccules, referred to as "wells" (Hermo et al. 1980; Ichikawa et al. 1984; Rambourg and Clermont 1986; Rambourg et al. 1984). The 50- to 80-nm vesicles are assumed to transport newly synthesized molecules from the nuclear envelope/endoplasmic reticulum to the Golgi apparatus and function in the transport between the individual cisternae of the stacks (e.g., Balch et al. 1984a, b; Farquhar and Palade 1981; Palade 1975; Rothman 1981). This "intercisternal transport system" includes the transfer of newly synthesized molecules between cisternae of different subsections of the stacks, presumably being cis-to-trans oriented, as well as trans-to-cis traffic supposedly playing a role in membrane reflux (Palade 1975; Rothman 1981) and in the transfer of Golgi products packaged at the cis side of the stacks (e.g., primary lysosomes, Brown and Farquhar 1984a). With some of the transport vesicles, a clathrin-negative cytoplasmic coat has been found associated; the latter vesicles have been considered as possibly being involved in "bulk" transport of newly synthesized molecules (Orci et al. 1986).

At the trans face of the stacks, the membranes of tansitional elements of the endoplasmic reticulum may be tightly associated with special transmost Golgi elements (Fig. 11; cf. Sect. 3.1.2).

Several reports have been published pointing to the existence of membrane continuities between endoplasmic reticulum and Golgi cisternae that may be significant for the transport of newly synthesized molecules (Bracker et al. 1971; Claude 1970; Flickinger 1973; Harris and Oparka 1983; Jaeken and Thines-Sempoux 1981; Kurosumi 1984; Mollenhauer and Morré 1976; Novikoff and Novikoff 1977; Ovtracht et al. 1973; Sasaki et al. 1984; Tanaka et al. 1986; Uchiyama 1982; Williams and Lafontaine 1985). Endoplasmic reticulum – Golgi interconnections include specialized smooth endoplasmic reticulum regions termed "boulevard peripherique" (Morré and Ovtracht 1981; Ovtracht et al. 1973) as well as specialized endoplasmic reticulum segments located at the trans Golgi face and termed GERL (Golgi-associated endoplasmic reticulum lysosomes, Novikoff 1964; cf. p. 23).

In arthropods and several types of vertebrate cells, beadlike structures, measuring 10 nm in diameter, are arranged in rings around the base of transport vesicles at the transitional elements of the endoplasmic reticulum (Brodie 1981,

1982a; Locke and Huie 1976a, b). The bead circles have been shown to collapse under the influence of inhibitors of oxidative phosphorylation (Brodie 1981), whereas inhibition of glycolysis and protein synthesis as well as disruption of the cytoskeletal system (Brodie 1981, 1982b) have failed to affect their arrangement. The integrity of the bead rings is believed to be essential for the transport of secretory materials from the endoplasmic reticulum to the Golgi complex. It has been suggested that the function of the bead circles is to specify a selected region of the rough endoplasmic reticulum where transition vesicles can form and that they prevent an intermixing with returning vesicles (Brodie 1982b).

Transitional elements have been found to increase in number and display long and complex forms upon arrest of endoplasmic reticulum-to-Golgi transport at low temperature and be essentially absent in the absence of ATP production (Tartakoff 1986).

3.1.2 Transmost Golgi Section

The transmost section of the Golgi stacks (referred to as the GERL-system by Novikoff 1964; rigid lamellae system by Claude 1970; trans-tubular network by Rambourg et al. 1979; trans Golgi reticulum by Willingham and Pastan 1984) is the main exit site for several newly synthesized molecules passing from the Golgi complex to other cellular locations. It has been shown that terminal steps in the processing of newly synthesized molecules, such as galactosylation and sialylation (Berger and Hesford 1985; Roth et al. 1984, 1985; Strous et al. 1983a) and propeptide-to-peptide conversions (Orci et al. 1985b), occur in transmost Golgi elements; it is here that a number of Golgi products are partitioned for different cellular destinations (e.g., Geuze et al. 1985; Griffiths et al. 1985). Transmost Golgi elements also function as recipients of endocytic materials (e.g., Gonatas et al. 1984; van Deurs et al. 1986; Figs. 9, 10) and have been shown to play a crucial role in the sorting of internalized ligands (for review, see Pastan and Willingham 1985).

The transmost Golgi subsection covers several structurally heterogeneous elements that are frequently, although not necessarily, located somewhat detached from the stack:

1. Flat or irregularly distended cisternae that possibly terminate in vacuolar segments, secretion granules (Fig. 4b)
2. Rigid lamellae (Figs. 6, 7a)
3. Tubular-reticular elements with numerous buds (Figs. 8, 9)
4. Polymorphous vacuoles possibly containing vesicular/cylindric inclusions and fingerlike extensions (Fig. 9)

The question as to whether these structures are different manifestations of one single compartment or represent individual entities has not yet been answered. Shape and arrangement of the trans Golgi elements are highly variable, being able to change in response to altered conditions (e.g., Griffiths et al. 1985; Hand and Oliver 1984b; Pavelka and Ellinger 1986a; Rambourg et al. 1981).

In common with the receptosome/CURL system (compartment of uncoupling receptor and ligand, Geuze et al. 1983a) and lysosomes, trans and trans-

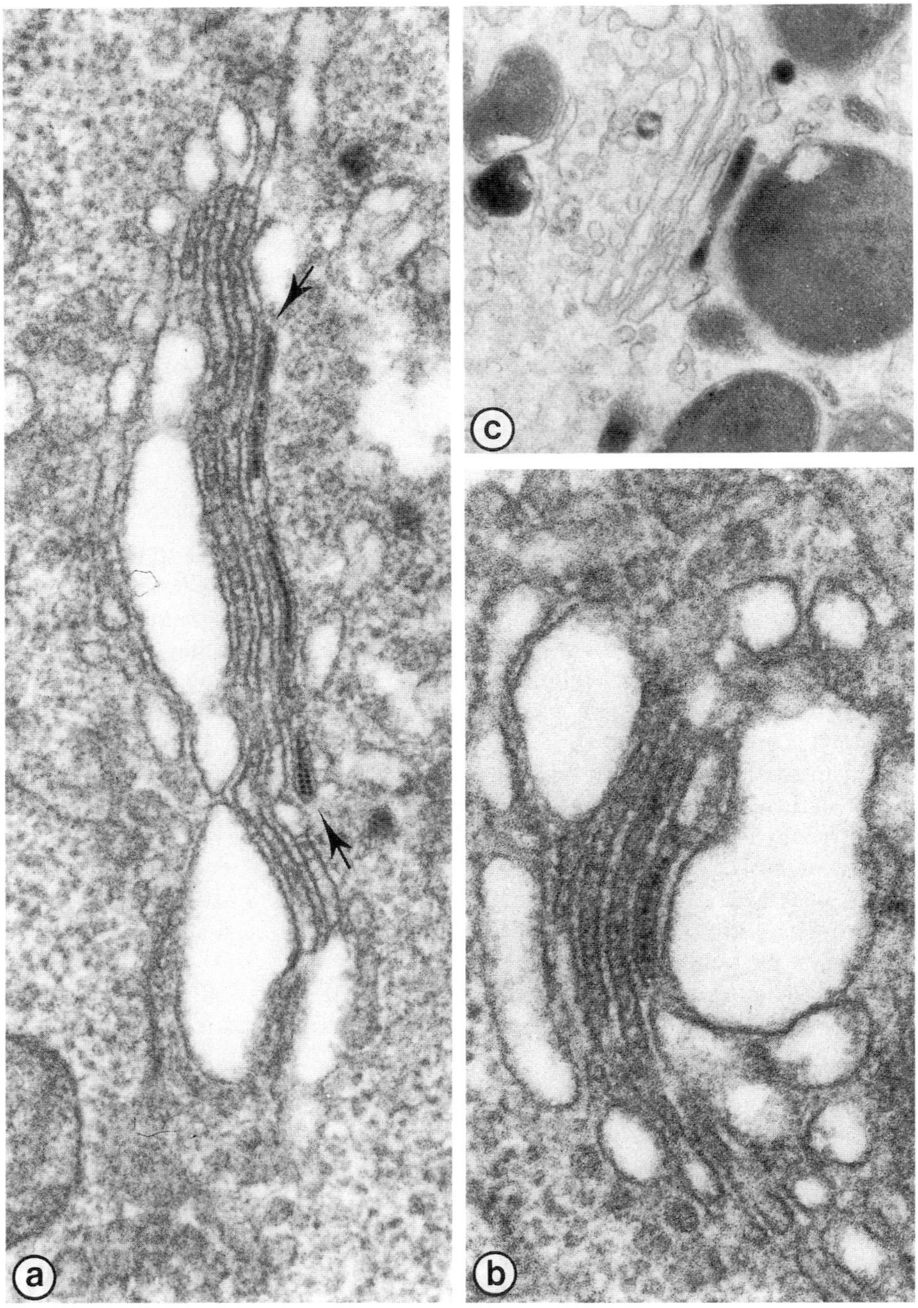

Fig. 10a–c. a, b Myeloma cells. Internalization of cationized ferritin, 60 min. **a** Internalized ferritin particles are apparent in vesicles and one cisterna (→) of the trans/transmost region of this stack. ×66000. **b** Medial cisternae of this stack are labeled by ferritin particles. ×80000. **c** Fibroblasts. Internalization of HRP-labeled PSA, 60 min. Reaction products, indicating internalized PSA-HRP conjugates, are localized in vesicles at both sides as well as in one cisterna (the transmost one?) of this stack. ×42500

most Golgi elements belong to the acidic cellular compartments (Anderson and Pathak 1985; Schwartz et al. 1985). Several recent studies have provided evidence for the existence in Golgi membranes of an ATP-dependent proton pump that accounts for acidification (Barr et al. 1984; Glickman et al. 1983; Zang and Schneider 1983). In fibroblasts, colocalization of fibronectin and dinitrophenol,3-(2,4-dinitroanilino)-3′-amino-N-methyldipropylamine (Anderson and Pathak 1985), the latter substance indicating low pH milieu, has demonstrated that acidic trans/transmost Golgi elements not only are involved in endocytic routes (cf. p. 21) but also are constituents of secretory pathways. It has been postulated (Anderson and Pathak 1985) that the pH milieu in Golgi elements may play a crucial role in the differential sorting of molecules (Rothman 1981) that traverse the Golgi apparatus. Molecules transported across the stacks by receptor-mediated processes, upon entering the low pH-trans elements, would dissociate from the receptors, which would permit the receptors to recycle back to medial/cis Golgi elements. Furthermore, acidic pH of the Golgi cisternae has been found to be required for multimerization of the von Willebrand factor (Wagner et al. 1986).

re 1, cisternae

The transmost cisternae may be flat or dilated, and they frequently exhibit cytoplasmically coated buds (Figs. 6a, 11b) and vacuolar distensions (Fig. 4b); the latter are particularly prominent in secreting cells and are considered as mirroring the pinching-off of secretion granules. In secretory cells, especially those of the "regulated" type, immature secretion granules (condensing vacuoles) are located close to the transmost Golgi cisternae (Figs. 4, 5); it is the compartment of the condensing vacuoles, which frequently exhibits extended clathrin coats, that has been shown to be the site of proteolytic propeptide-to-peptide conversion in the endocrine pancreatic cells (Orci et al. 1985b).

re 2, rigid lamellae

The transmost cisternae may exhibit conspicuously uniform luminal spaces (20–30 nm), "thicker" membranes as compared with the stacked saccules, and central dense lines ("midline densities": Figs. 6, 7); these structures have been termed "rigid lamellae" (Claude 1970, Hand and Oliver 1984a). Knoblike structures and balloonlike distensions of the rims indicate vesicle budding and the formation of secretory granules, which suggests involvement of the rigid lamellae in the secretory pathway. However, secretory molecules have rarely been demonstrated in the luminal spaces of rigid lamellae (Bendayan 1984; Broadwell et al. 1979; Geuze et al. 1979; Hand and Oliver 1977a, b, 1981; Orci et al. 1984a).

The rigid lamellae system is particularly extended in the early stages of cell differentiation (Figs. 6, 7a, 11b; Pavelka and Ellinger 1986a), then being closely associated with segments of the endoplasmic reticulum.

It is noteworthy that rigid lamellae are not limited to the trans Golgi region; in some cell types, such as in the small intestinal epithelial cells of the crypt-top region, "rigid" cisternae are located in medial subsections of the Golgi stacks (Fig. 7b).

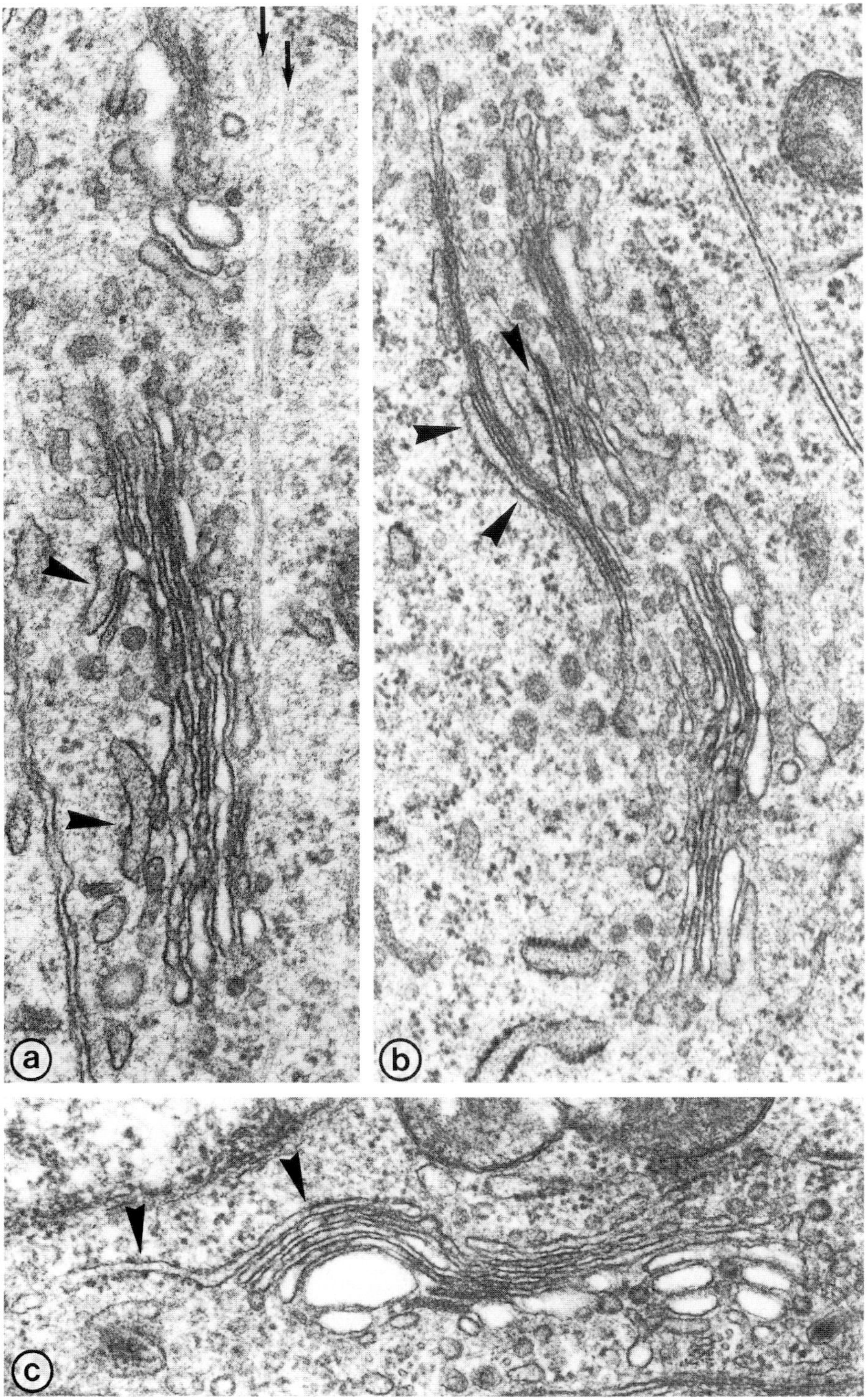

Fig. 11 a–c

re 3, tubular-reticular elements

In a variety of cell types, the transmost cisternae continue in branched tubular elements that may constitute a prominent tubular-reticular system at the trans side of the stacks (Figs. 8, 9). Numerous knoblike protrusions, frequently with cytoplasmic coats, indicate vesicle budding. Two kinds of coated buds have been distinguished (Griffiths et al. 1985; Orci et al. 1986), some of them containing clathrin, others being devoid of immunocytochemical staining with anti-clathrin antibody.

Several findings indicate that the branched tubular elements at the trans Golgi side are sites of terminal glycosylation (Geuze et al. 1985; Roth et al. 1985) and exit sites of newly synthesized membrane proteins (Griffiths et al. 1985), secretory molecules (e.g., Geuze et al. 1985), and lysosomal enzymes (Geuze et al. 1985; Novikoff 1976). Furthermore, tubular structures of the transmost Golgi section function as recipients and sites of partitioning of internalized molecules (for review, see Pastan and Willingham 1985; cf. also pp. 22 and 56).

re 4, polymorphous vesicular bodies

In addition to cisternae and tubules, polymorphous vesicles with tubular appendices and vesicular/cylindric inclusions, reminiscent of multivesiculated bodies, are prominent trans Golgi elements (e.g., Figs. 9a–c). Similar to the findings at tubular-reticular structures, polymorphous vesicles have been found to be part of the pathway of newly synthesized molecules (Griffiths et al. 1985); on the other hand, they may contain endocytic materials and are reminiscent of the receptosome/CURL system. The latter compartment, known by the terms "endosome" (Helenius et al. 1983; Hopkins 1985), "receptosome" (Willingham and Pastan 1980; 1985b), "CURL" (compartment of uncoupling receptor and ligand, Geuze et al. 1983a), and "pale multivesiculated body" (Paavola et al. 1985), is considered to be the site of the dissociation and segregation of receptors and ligands in receptor-mediated endocytosis. It appears to play a crucial role in the intracellular routing of internalized ligands and receptors (e.g., Geuze et al. 1983a, b, 1984b; Paavola et al. 1985; Pastan and Willingham 1981; Willingham et al. 1983a, 1984; Wolkoff et al. 1984) as well as in the trafficking of newly synthesized molecules such as lysosomal enzymes (Geuze et al. 1984a). A similar sorting compartment, in which segregation of endocytic materials and membrane constituents occurs, has been described for fluid phase endocytosis (Storrie et al. 1984). Elements of the Golgi apparatus, e.g., the polymorphous

◁ **Fig. 11a–c.** "Golgi-associated endoplasmic reticulum." Golgi stacks of epithelial cells in early stages of cell differentiation: embryonic pancreatic acinar cells, days 15/16 of gestation (**a**, **b**), and crypt cells of the small intestine (**c**). **a** The cisternae are irregularly dilated; at one side, segments of endoplasmic reticulum (▶) are closely apposed to the Golgi saccules. Microtubules (→) run in parallel with the Golgi cisternae. ×50000. **b** Cisternae of the endoplasmic reticulum with ribosomes attached to one surface only (=transitional elements) are located in tight association with cisternae of the trans side of this stack; the endoplasmic reticulum cisternae (▶) occupy the transmost position and are interposed between individual trans Golgi saccules ×45000. **c** A cisterna of the endoplasmic reticulum (▶) runs in parallel and close spatial relationship to a trans saccule of this Golgi stack. ×42500 vesicular-tubular structures in the transmost position, possibly function as elements of uncoupling of ligand and receptor similar to CURL (Breitfeld et al. 1985; Geuze et al. 1985). The finding that trans/transmost Golgi elements possess acidic milieu supports such a role.

3.1.3 Golgi Elements – "Common Secretory-Endocytic Compartments"?

Uptake of insulin into elements of the Golgi apparatus has been studied in hepatocytes (Kay et al. 1984); the majority of internalized insulin has been found in structures being devoid of galactosyltransferase.

On the other hand, diverse common characteristics, including morphologic similarity (Geuze et al. 1985), low pH, and similar effects of lysosomotropic agents (Anderson and Pathak 1985; Schwartz et al. 1985; Strous et al. 1985a) as well as the results obtained with a mutant Chinese hamster ovary cell line (Robbins et al. 1984) suggest a close relationship between elements of the trans Golgi side involved in the terminal processing of newly synthesized molecules and parts of the endocytic system (cf. also Sect. 5).

Both tubular-reticular and multivesiculated trans Golgi elements have been found in connection with newly synthesized molecules, on the one hand, and endocytic substances, on the other hand. Trans Golgi tubular-reticular elements and multivesiculated structures are serious candidates for a possibly existing common secretory-endocytic compartment that might play an essential role in the restoration and targeting of internalized molecules. For example, it has been shown that internalized asialotransferrin (Regoeczi et al. 1982) and asialotransferrin receptor (Snider and Rogers 1985) become resialylated; resialylation might take place in trans Golgi tubular-reticular elements, which are known to be recipients of internalized transferrin (Willingham et al. 1984; Yamashiro et al. 1984) and rich in sialyltransferase (Berger and Hesford 1985; Roth et al. 1985).

Various endocytic substances as well as recycling plasma membrane proteins not only gain access to trans/transmost Golgi elements, but also visit medial and cis cisternae of the Golgi stacks (e.g., Herzog and Farquhar 1977; Snider and Rogers 1986; Woods et al. 1986); hence, in the sense of definition of common compartments for newly synthesized and internalized molecules, the entire Golgi apparatus can be seen as a "common secretory-endocytic compartment." Here, the routes of newly synthesized, endocytic, and recycling molecules cross. Newly synthesized as well as endocytic and recycling molecules are subjected to certain modifications and, subsequently, are targeted to diverse cellular locations.

3.1.4 Endoplasmic Reticulum

Cisternae of the endoplasmic reticulum are constant elements of the transmost Golgi section (Fig. 11; Broadwell and Oliver 1983; Cataldo and Broadwell 1982; Hand 1980; Hermo et al. 1979; Novikoff and Novikoff 1977; Pavelka and Ellinger 1983b, 1986a; Rambourg et al. 1981). In many instances, they represent transitional elements, polyribosomes being confined to one side of the cisternae.

Endoplasmic reticulum segments frequently are located close to transmost Golgi elements, e.g., they run in parallel with rigid lamellae or are interposed between individual trans Golgi cisternae (Fig. 11b; Broadwell and Oliver 1983; Cataldo and Broadwell 1982; Hermo et al. 1979; Pavelka and Ellinger 1983b, 1986a). Such tight trans Golgi-endoplasmic reticulum associations are particularly prominent in early stages of cell differentiation, such as in the early developmental period of the embryonic pancreatic acinar cells (Fig. 11b; Pavelka and Ellinger 1986a) and in immature epithelial cells of the small intestinal crypts (Fig. 11c; Pavelka and Ellinger 1983b). Despite of the close and wide-spaced associations between the two compartments, in the latter studies membrane continuities between endoplasmic reticulum segments and Golgi elements were not demonstrable.

3.1.5 Golgi-Associated Endoplasmic Reticulum Lysosomes

Most of the trans Golgi structures described in the preceding paragraphs, i.e., transmost Golgi cisternae, rigid lamellae, condensing vacuoles, trans Golgi tubular-reticular structures, are part of a system that was presented by Novikoff in 1964 (Novikoff 1964) and given the name GERL (Golgi-associated endoplasmic reticulum lysosomes). GERL denotes a specialized region of the endoplasmic reticulum (ER) located at the trans face of the Golgi stacks (G), which plays a part in the formation of lysosomes (L). In the ensuing years, GERL was described in many cell types, the original definition having been extended. It was shown that GERL elements are connected with the packaging of secretory materials (e.g., Novikoff et al. 1977; Hand and Oliver 1977a, b; for review Hand and Oliver 1984a) and function as recipients of endocytic molecules (Gonatas et al. 1977; Haimes et al. 1981; Novikoff et al. 1981). Most essentially, the GERL concept suggests interconnections between rough endoplasmic reticulum and elements of the trans Golgi side (trans cisternae, condensing vacuoles). It has been held that such interconnections provide a direct endoplasmic reticulum-to-trans Golgi route for newly synthesized secretory and lysosomal molecules, by which the stacked Golgi cisternae may be bypassed (Novikoff et al. 1975, 1977).

Subsequent studies have shown that a strict differentiation between GERL and Golgi apparatus is not justified (e.g., Broadwell and Oliver 1981; Goldfischer 1982; Hand 1980; Hand and Oliver 1984b; Oliver and Hand 1983; Oliver et al. 1980; Pavelka and Ellinger 1982, 1983b; Sawano and Fujita 1981). Functional processes characteristic of the Golgi apparatus, e.g., terminal glycosylation, have been found associated with both cisternae of the stacks and transmost Golgi elements (e.g., Roth et al. 1985). Secretory and lysosomal molecules have been localized immunocytochemically in the stacked Golgi cisternae (among many others, Bendayan 1984; Slot and Geuze 1983; van Dongen et al. 1984), an argument against a route that bypasses the Golgi stacks. Nevertheless, the classic publications on GERL, viz., those by Novikoff and co-workers (Holtzman et al. 1967; Novikoff 1976; Novikoff and Novikoff 1977; Novikoff et al. 1968, 1975, 1977; Novikoff and Yam 1978; Novikoff et al. 1971) and by Hand and Oliver (Hand 1971, 1980; Hand and Oliver 1977a, b, 1984a), are minute descriptions of the trans face of the Golgi stacks, representing bases for further morphologic and cytochemical work on Golgi architecture. Essential contents of the GERL concept have been proved by recent studies; thus, it has been shown that transmost Golgi elements, formerly included in the term GERL, are the main sites of the packaging of several Golgi products, including secretory and lysosomal molecules, before leaving the Golgi complex for other cellular destinations.

3.1.6 Cytoplasmically Coated Membranes

A variable number of cytoplasmically coated membrane profiles (coated buds/coated vesicles?) are found close to cis as well as trans Golgi cisternae (Figs. 4, 6, 7; Croze et al. 1982; Friend and Farquhar 1967; Geuze and Kramer 1974; Jamieson and Palade 1967a, b, 1971; Kartenbeck et al. 1981; Merisco et al. 1982; Novikoff 1976; Willingham et al. 1981a). They are considered as being involved in the intracellular routing of endocytic molecules (Pastan and Willingham 1985; Willingham and Pastan 1984) as well as newly synthesized membrane glycoproteins (Rothman and Fine 1980) and secretory molecules (Franke et al. 1976; Goldenberg and Fine 1985). Cytoplasmic coats are observed at membranes of each of the Golgi subsections; they are particularly prominent at elements of the trans Golgi side, here mostly being associated with knoblike protuberances. Clathrin (Pearse 1976) has been localized immunocytochemically at coated Golgi membrane segments (Orci et al. 1984b, 1985a; Willingham et al. 1981a), although, in a recent study, several of the coated buds did not stain with the anticlathrin antibody (Griffiths et al. 1985). This finding, as well as others obtained with a cellfree system that reconstitutes transport between successive cisternal compartments of the Golgi stack (Orci et al. 1986), indicates the existence of different classes of coated Golgi membrane segments and coated vesicles. Evidence has been provided for clathrin-negative-coated vesicles as being involved in protein transport across the stack of cisternae, possibly functioning as bulk membrane carriers (Orci et al. 1986).

In colocalization with clathrin, 100-kDa proteins of the cytoplasmic coat, supposedly important for binding the coat to the membrane, have been immunocytochemically demonstrated in the Golgi area (Robinson and Pearse 1986).

Receptor-mediated transport to diverse cellular locations of Golgi products (e.g., Brown et al. 1984; Geuze et al. 1985; Novikoff and Novikoff 1977), endocytic materials (Hanover et al. 1984; Willingham and Pastan 1982; Willingham et al. 1981a) and molecules retrieved from the plasma membrane (Beguinot et al. 1984) appear to be connected with the clathrin-coated Golgi membrane segments. Coated buds are considered as being related to the formation of primary lysosomes (Brown et al. 1984) and some of the vesicles involved in the transfer from the Golgi complex to other cellular compartments, e.g., in the Golgi-to-plasma membrane transport of newly synthesized molecules and the Golgi-to-lysosome route of internalized ligands (e.g., Hanover et al. 1984).

In endocrine pancreatic cells, a clathrin-coated compartment in the transmost Golgi position has been shown to be the site of proinsulin-to-insulin conversion (Orci et al. 1985b). It is this compartment, in which nonconverted, amino acid analog-modified proinsulin is arrested (Orci et al. 1984c) and proinsulin is accumulated upon presence of monensin (Orci et al. 1984b).

In pancreatic acinar cells, cytoplasmically coated membrane profiles have been found to be overabundant in the trans Golgi area following treatment with the antimicrotubular drug colchicine (Pavelka and Ellinger 1983a) and in the absence of ATP production (Tartakoff 1986). The latter finding may be connected with ATP requirement for the removal of clathrin from coated membranes (Braell et al. 1984b; Schlossman et al. 1984; Schmid et al. 1984).

As a consequence of treatment with the N-glycosylation-inhibitor tunicamycin, coated vesicles showing immunoreaction for mannose-6-phosphate receptor have been found accumulated near the cis Golgi face (Brown et al. 1984). This finding is considered as reflecting arrest of Golgi-to-lysosome traffic via coated vesicles due to lack of lysosomal enzymes containing the mannose-6-phosphate recognition marker. Similar accumulation of coated vesicles has been observed in I-cell fibroblasts from patients with mucolipidosis II, which are also deficient in mannose-6-phosphate-containing ligands (Brown and Farquhar 1984b).

3.2 Golgi Architecture During Mitosis

During mitosis, the Golgi apparatus undergoes massive alterations (e.g., Melmed et al. 1973; Schroeter et al. 1985; Zeligs and Wollman 1979). In the late prophase, the location of the Golgi apparatus relative to the nucleus is no longer maintained, the stacks being dispersed in the entire cytoplasm. In the prometaphase to metaphase stages, the Golgi complex appears less and less prominent; the saccules are shortened and distended; they are accompanied by an increased number of small 60- to 70-nm vesicles, reminiscent of transport vesicles. In the metaphase and anaphase, the stacked appearance of the Golgi cisternae is disturbed; prominent accumulations of small (60–70 nm) vesicles are intermixed with vacuolar remnants of the Golgi stacks. In the telophase, the cisternae recover their flat, elongated conformation; the stacks are reformed and the vesicle accumulations disappear. Despite the massive morphologic alterations, the Golgi elements retain cytochemical staining characteristics (Schroeter et al. 1985).

Galactosyltransferase has been localized immunocytochemically in the dispersed Golgi vacuoles (Hiller and Weber 1982). The vacuolized cisternae also stain with an antibody directed against a Golgi-associated protein (Burke et al. 1982; Schroeter et al. 1984); Golgi-derived vacuoles, reacting with the latter antibody, have been demonstrated around the forming spindle. In the late anaphase to telophase, they have been found penetrating the spindle zone, then being located close to the telomere region of the chromosomes (Schroeter et al. 1984).

During mitosis, intracellular traffic of newly synthesized as well as internalized molecules is impaired (for review, see Warren 1985). Using G protein of the vesicular stomatitis virus as a model, it has been shown that the transport of membrane proteins out of the endoplasmic reticulum to the Golgi complex is inhibited (Featherstone et al. 1985). The disintegration of the Golgi apparatus during mitosis has been considered as possibly being a consequence of impaired endoplasmic reticulum-to-Golgi traffic.

A model recently presented (Warren 1985) has suggested that intracellular traffic stops because of an inhibition of vesicle fusion with vesicle budding

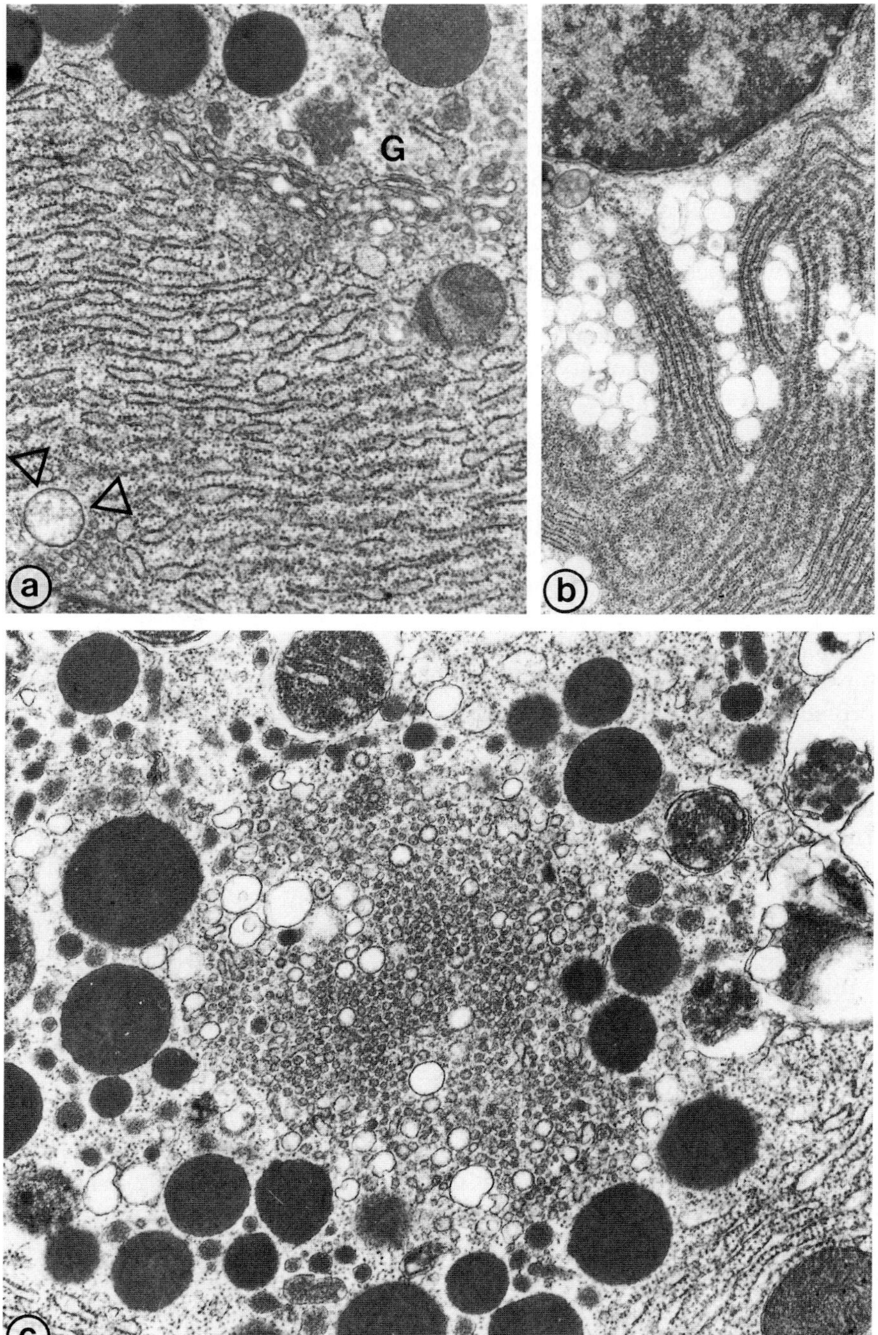

Fig. 12a–c. Colchicine treatment. **a** Pancreatic acinar cell of the rat, colchicine 30 min. The stacked Golgi cisternae (*G*) appear unaltered; accumulations of small vesicles, closely apposed to larger vacuoles (▷), are found outside the virtual Golgi area. ×20000. **b** Rat duodenal goblet cell, colchicine 45 min. Prominent accumulations of vacuoles characterize the basal cytoplasmic areas. ×13000. **c** Rat pancreatic acinar cell, colchicine 3 h. Stacks of cisternae are missing; the Golgi area is characterized by extended accumulations of small vesicles, vacuoles, and small secretory granules. ×23000

continuing, which leads to a fragmentation of nuclear envelope/endoplasmic reticulum and Golgi stacks.

The Golgi apparatus alterations during mitosis strikingly resemble the disorganization of this organelle under the influence of microtubule-disrupting agents (Fig. 12). Treatment with, e.g., colchicine, results in displacement of the Golgi stacks, vacuolization and disruption of the stacked character of the Golgi saccules, as well as massive accumulation of small, presumably transport, vesicles. It seems likely that both the alterations of the Golgi complex during mitosis as well as after treatment with microtubule-affecting agents are related to an altered organization of the microtubule system.

3.3 Microtubules and Golgi Organization

Microtubules are abundant in the Golgi area (Figs. 7, 11) and have been found associated with elements of the Golgi apparatus (Rogalski and Singer 1984), sometimes forming regular arrays in the vicinity of the Golgi saccules (Fig. 11a). In many cell types, Golgi apparatus and microtubule-organizing center have been localized in close proximity (e.g., DeCamilli et al. 1986; Kupfer et al. 1982, 1983) and have been shown to redistribute together upon altered functional conditions of the cells.

The significance of the microtubular system in regard to Golgi organization becomes evident by studies using agents that interfere with microtubule function, such as colchicine and *Vinca* alkaloids, which disrupt the microtubule system, or taxol, which induces polymerization of microtubules without relation to the microtubule-organizing centers. Such agents cause disorganization of the Golgi complex (e.g., Bennett et al. 1984a, b; Blok et al. 1981; Busson-Mabillot et al. 1982; Cho and Garant 1981b; Glickman et al. 1976; Knudson et al. 1978; Magalhaes et al. 1985; Patzelt et al. 1977; Pavelka and Ellinger 1981b, 1983a; Pavelka and Gangl 1983; Pavelka et al. 1983; Reaven and Reaven 1977; Redman et al. 1975; Rindler et al. 1984b; Rogalski et al. 1984; Sandoval et al. 1984; Thyberg and Moskalewski 1985; Thyberg et al. 1980; Wehland and Sandoval 1983; Wehland et al. 1983; Williams 1981).

The results of studies by the author with pancreatic acinar cells (Pavelka and Ellinger 1983a), serous cells of the submandibular gland, and goblet cells are briefly summarized in the following. The earliest colchicine-induced alterations were observed 20–30 min after application of the drug and concerned the enzyme-cytochemical patterns: TPPase, IDPase, and AcPase activities were decreased. At the same time, vacuoles and "Golgi-like" formations occurred outside the virtual Golgi area (Fig. 12a, b). In later stages, the organization of the Golgi saccules was disrupted; after periods of 90 min after colchicine treatment and later, regular stacks of cisternae were hardly discernible. The cisternae were vacuolized, groups of vacuoles and extensive accumulations of small vesicles, reminiscent of transport vesicles, characterized the Golgi area (Fig. 12c), and Golgi elements appeared dispersed throughout the entire cytoplasm.

Recent studies have shown that Golgi architecture is maintained by microtubules organized by the perinuclear microtubule-organizing center(s). Agents inducing polymerization of microtubules without relation to the microtubule-organizing centers, such as taxol (Sandoval et al. 1984; Wehland et al. 1983) and guanosine 5′-(α,β-methylene)triphosphate (Wehland and Sandoval 1983), perturb the organization of Golgi elements. After treatment with taxol, which causes polymerization of microtubules predominantly located in the cell periphery, the Golgi stacks appear fragmented, interspersed with microtubules, and arrayed in the vicinity of the peripheral bands of microtubules (Wehland et al. 1983).

A close connection of Golgi apparatus and microtubular system is also indicated by the immunocytochemical localization of cyclic-AMP-dependent protein kinase type II in various cell types: regulatory and catalytic subunits have been found concentrated in the Golgi region and associated with microtubules and microtubule-organizing center (DeCamilli et al. 1986; Nigg et al. 1985a). This may signalize functional relationship of these organelles. It has been considered that cAMP-dependent protein kinase activity may be active in mediating processes involved in microtubule organization and function, phosphorylation of microtubule-associated protein II possibly playing a role (DeCamilli et al. 1986; Nigg et al. 1985a).

The fact that Golgi architecture is influenced by the microtubular system has also been demonstrated elegantly by microinjection of anti-α-tubulin antibodies; intracellular injection at high concentrations has been shown to induce perinuclear aggregation of microtubules and dispersion of Golgi elements, analogous to the findings after colchicine treatment (Wehland and Willingham 1983; Wehland et al. 1983).

Regular organization and location of the Golgi apparatus appear to be essential factors for normal Golgi-to-cell surface transport of plasma membrane constituents and for the maintenance of the polar cell organization. Disruption of the Golgi apparatus by means of microtubule-perturbing agents seems to be related to altered cell surface characteristics and disturbance of cell polarity occurring under the same conditions (Berlin et al. 1979; Buschman 1983; Danielsen et al. 1983; Ellinger et al. 1983; Oliver and Berlin 1982; Pavelka et al. 1983; Quaroni et al. 1979; Rindler et al. 1984b; Rogalski et al. 1984). Details of the colchicine-induced cellular disorganizations in small intestinal absorptive cells were studied in our laboratory; colchicine caused dislocation, vacuolization, and altered cytochemical staining of the stacked Golgi cisternae (Pavelka and Ellinger 1981b; Pavelka et al. 1983). The Golgi alterations preceded alterations of the cell polarity that became evident by the occurrence of microvillus borders, reminiscent of the apical brush border, at the lateral and basal cell surfaces (Fig. 13); at 6 h after the administration of colchicine, brush borders occupied an average of 6% of the basolateral cell surface of the absorptive cells. The altered cell polarity was also demonstrated by radioautographic studies with radiolabeled fucose; this sugar is inserted in newly synthesized glycoproteins

Fig. 13a–c. Colchicine treatment. **a**, **b** Radioautography, 4 h following intraperitoneal injection of 3H-labeled fucose. **a** Control. Intense silber grain label is apparent at the brush border lining the apical surface of the epithelial cells; the massive brush border label contrasts to the weakly stained basolateral surface regions. The Golgi area is moderately labeled. ×975. **b** Colchicine treatment, 6 h. Intense silber grain label lines apical as well as basolateral surface regions of the epithelial cells. ×950. **c** Colchicine treatment, 6 h. A microvillus border (———), reminiscent of the apical brush border, is apparent at the lower lateral cell surface of this absorptive cell. *bm*, basement membrane; ➛ cross sections of microvilli. ×24000 ▷

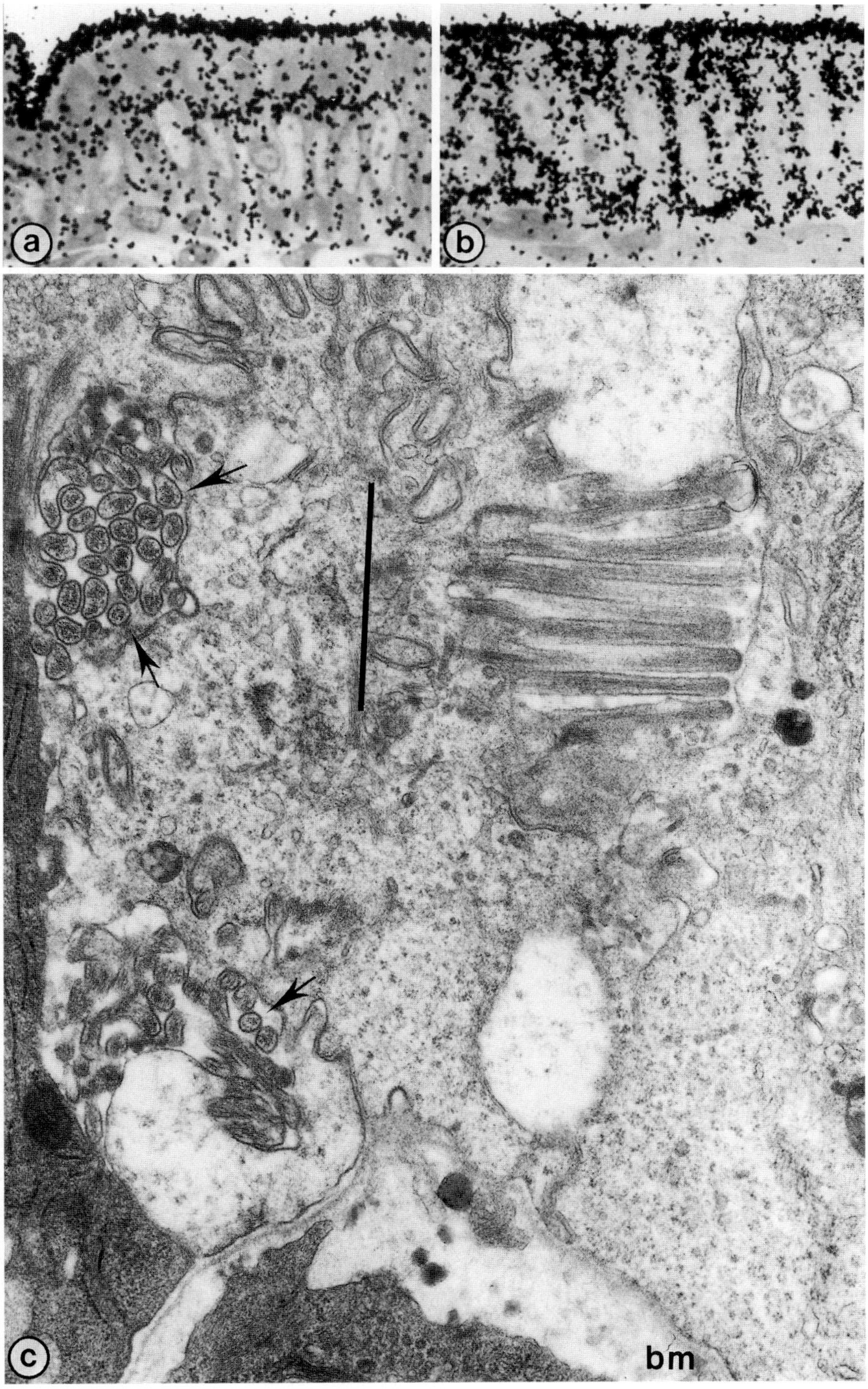

Fig. 13a–c

at the Golgi apparatus level (Bennett and Leblond 1977). In controls, the subsequent redistribution and incorporation in the plasma membrane of the glycoproteins cause intense radioautographic label over the apical cell surface and slight reaction over the basolateral surface regions. This clear difference disappeared after treatment with colchicine: the label intensity at the basolateral cell surfaces was equal to that at the apical cell surface (Fig. 13a, b; Ellinger et al. 1983). These findings are in line with recent biochemical results obtained with absorptive cells (Quaroni et al. 1979) as well as findings on the insertion into plasma membrane of virus glycoproteins (Rindler et al. 1984b; Rogalski et al. 1984). It appears that disruption of microtubules particularly impairs Golgi domains that are essential for the targeting of Golgi products, e.g., of plasma membrane glycoproteins, to their proper destinations, whereas other functions, e.g., terminal glycosylation, are unaffected (Banerjee et al. 1976; Rogalski et al. 1984).

Another colchicine effect may also be related to the Golgi apparatus and/or plasma membrane alterations: in the absorptive cells (Blok et al. 1981; Buschman 1983; Ellinger and Pavelka 1984b), just as in some other cell types (Michaels 1983; Thyberg et al. 1982), colchicine induced extensive accumulations of tubular-vesicular-cisternal organelles. These were reactive for diverse phosphatases (TPPase, IDPase, AcPase, TMPase) and recipients of endocytic materials taken up at the apical and basolateral cell surfaces (Ellinger and Pavelka 1986). According to these findings, it is likely that the accumulated organelles belong to the endosomal and/or lysosomal compartments. Accumulation of such organelles suggests that colchicine impairs endocytic pathways; these may include endosome-to-lysosome (cf. Wolkoff et al. 1984), endosome-to-Golgi (cf. Thyberg and Stenseth 1981), and endosome-to-plasma membrane (recycling) routes.

Intercisternal Elements

In some kinds of plant cells, arrays of thin 7- to 8-nm fibers are apparent between the cisternae of the Golgi stacks (Mollenhauer 1965), and intercisternal material has been demonstrated in the Golgi complex of *Trichomonas* (Amos and Grimstone 1986).

Cytoplasmically oriented domains of transmembrane proteins of the Golgi cisternae are thought to play a role in membrane-membrane interactions that may be essential for the preservation of the parallel arrangement of the saccules. The close association with each other of the Golgi saccules is preserved even under conditions that massively perturb Golgi architecture, such as treatment with the K^+/Na^+-ionophore monesin, inducing extreme dilatation of the Golgi saccules (e.g., Ellinger and Pavelka 1984a; Tartakoff 1983b). Recently, two cytoplasmically oriented Golgi membrane proteins that may have a function in membrane-membrane interactions have been localized immunocytochemically at medial saccules of the Golgi stacks (Chicheportiche et al. 1984).

3.4 Variability of Golgi Architecture

The batterylike arrangement of sets of cisternae, which build up the characteristic stacks, is a basic feature of the Golgi apparatus that accounts for its typical architecture and applies to each cell type.

However, many architectural details of the Golgi organelle are variable; morphologic characteristics and cytochemical patterns, covering enzyme-cytochemical, immunocytochemical, and lectin-cytochemical reactions, may considerably differ between diversely specialized cells and correspondingly change with altered cell function (e.g., Blomfield et al. 1983; Bogard 1975; Broadwell and Oliver 1981; Doine et al. 1984; Ellinger and Pavelka 1982; Flickinger 1978; Friedman and Cardell 1976; Griffiths et al. 1985; Hand and Oliver 1984b; Hickey et al. 1983; Jamieson and Palade 1971; Orci 1982; Paavola 1978a, b; Pavelka and Ellinger 1982, 1986a; Romagnoli 1984; Scott and Flickinger 1983; Sasaki et al. 1984; Slot and Geuze 1979; Susi et al. 1971; Tang et al. 1982; Treilhou-Lahille 1982; Uchiyama and Saito 1982; Völkl et al. 1976; Weakley et al. 1981; for a recent review, see Oliver and Hand 1983). This particularly concerns the number and shape of the cisternae of the individual stacks, the organization of the transmost Golgi elements and their relationship to the endoplasmic reticulum, the location of clathrin coats, the enzyme-cytochemical patterns, and the distribution in Golgi elements of lectin binding sites. The question as to whether these variabilities mean function-related variations of certain domains of the Golgi apparatus or reflect changes of its overall organization has not yet been answered.

Changes of the Golgi apparatus have been shown under conditions that specifically stimulate certain cell functions; they particularly concern enzyme-cytochemical reactions (reviewed in Oliver and Hand 1983; cf. p. 37). In another approach, variations of the Golgi architecture have been observed in the course of cytodifferentiation (Burchanowski et al. 1982; Flickinger 1978; Hickey et al. 1983; Sasaki et al. 1984; Sterle and Pipan 1985; Susi et al. 1971; Tang et al. 1982; Treilhou-Lahille 1982; Weakly et al. 1981). The author's studies concentrated on the architecture of the Golgi apparatus in the course of cell differentiation of rat embryonic acinar cells and small intestinal absorptive cells (Ellinger and Pavelka 1982; Pavelka and Ellinger 1983b, 1986a); they revealed variations that particularly concerned the transmost Golgi elements.

3.4.1 Golgi Architecture in the Course of Cell Differentiation

Rigid Lamellae System

In the protodifferentiated and early differentiated states of pancreatic acinar cells (Pictet et al. 1972), the system of rigid lamellae was extensively developed (Figs. 6, 7a), exhibiting numerous buds; in the later stages of development, rigid lamellae were less prominent; except for a few short segments, they were absent in the prenatal period. By contrast, in the small intestinal absorptive cells, the transmost Golgi cisternae rarely exhibited features of rigid lamellae. On the other hand, cisternae with a constant luminal width and "midline densities," reminiscent of rigid lamellae, were located in medial subsections of the Golgi stacks (Fig. 7b); this was primarily observed in the epithelial cells of the crypt-top region.

Endoplasmic Reticulum-Trans Golgi Associations

Early stages of differentiation of both embryonic pancreatic acinar cells (Fig. 11b) and absorptive cells (Fig. 11c) were characterized by tight associations of transmost Golgi cisternae and segments of the endoplasmic reticulum. Transitional elements of the endoplasmic reticulum, with ribosomes attached to one surface only, ran in parallel and closely adjacent to the transmost Golgi cisternae. Endoplasmic reticulum mostly occupied the ultimate position at the trans side of the stacks (Fig. 11 c), although it was also found interposed between individual trans Golgi cisternae (Fig. 11 b), thus being an integral element of the stacks.

In both cell types, the tight endoplasmic reticulum-trans Golgi relationship became less apparent as the cells matured. The functional significance of such close and wide-spaced endoplasmic reticulum-Golgi associations has not yet been clarified; the predominence in early stages of cell differentiation suggests that the close relationship between the two compartments may in some way be connected with the formation of trans Golgi elements.

4 Cytochemistry

4.1 Classic Cytochemical Reactions – Enzyme Cytochemistry

The classic cytochemical Golgi reactions include "metallic impregnation techniques" (e.g., reduction of silver salts, prolonged osmification, osmium iodide reactions, Friend and Murray 1965; Locke and Huie 1983; lead staining, McClintock and Locke 1982), the enzyme-cytochemical localization of TPPase, NDPase, and AcPase (e.g., Barka 1964; Novikoff and Goldfischer 1961), as well as cytochemical techniques for localizing complex carbohydrates (high iron diamine, periodic acid silver techniques, Rambourg et al. 1969; Spicer and Schulte 1982; lectin cytochemistry, pp. 42–55). Cytochemical procedures for demonstration of nicotinamide adenine dinucleotide phosphatase (NADPase) have recently been described (Smith 1980, 1981) and employed at several cell types (e.g., Clermont et al. 1981; Parsons and Smith 1984).

The mechanisms underlying the classic metallic staining techniques may be closely related to the reactions by which Camillo Golgi discovered the "apparato reticulare interno" that bears his name. It has been suggested that the reactions obtained by classic osmium methods localize labile S-S bridges which are able to convert to SH (Locke and Huie 1983).

The enzyme-cytochemical TPPase, NDPase, and AcPase reactions have been thought to be related to the breakdown of nucleoside phosphates originating in the course of glycosylation (Kuhn and White 1977; Yamazaki and Hayaishi 1968). In HeLa cells, TPPase (corresponding to uridine diphosphatase, UDPase) has been colocalized with the immunocytochemical reactions of galactosyltransferase (Roth and Berger 1982). In analogy, AcPase has been considered as being involved in the breakdown of cytidine monophosphate, originating in connection with sialic acid transfer (Berger 1984; Novikoff et al. 1971). It is uncertain in what respect this is functionally significant. It has recently been shown that the Golgi apparatus membranes contain specific carrier proteins for sugar nucleotides, just as for nucleotide sulfate (Capasso and Hirschberg 1984; Deutscher and Hirschberg 1986; Perez and Hirschberg 1985; Schwarz et al. 1984). Translocation of sugar nucleotides takes place via a coupled equimolar exchange with the corresponding nucleoside monophosphates (Capasso and Hirschberg 1984).

Furthermore, it is questionable whether the cytochemical AcPase reactions obtained by using β-glycerophosphate as substrate indicate locations of lysosomal enzymes or reflect other enzyme activities, such as enzymes engaged in the cleavage of newly synthesized molecules. In secretory cells of the lateral prostate, secretory AcPase reacting with naphthol-As-Bi-phosphate as substrate has been distinguished cytochemically from lysosomal AcPase, reacting with β-glycerophosphate (Kimura and Ichihara 1985). The secretory enzyme has been found in Golgi saccules, condensing vacuoles, and secretory vesicles and lysosomal AcPase in the transmost (GERL-like) elements of the Golgi complex and in lysosomes.

The location of the lysosomal arylsulfatase in the set of stacked Golgi cisternae (Baron et al. 1985; Quatacker 1979) is in accordance with the immunocytochemical demonstration of a series of lysosomal enzymes in all the stacked Golgi saccules of fibroblasts and hepatocytes (cf. p. 39; Geuze et al. 1985; Van Dongen et al. 1984).

Furthermore, alkaline phosphatase (e.g., Paavola 1978a, b), pyridoxal phosphatase (Spater et al. 1978), 5′-nucleotidase (e.g., Farquhar et al. 1974; Uchiyama 1983; Widnell et al. 1982), glucose-6-phosphatase (Broadwell and Cataldo 1983; Cataldo and Broadwell 1984), tyrosinase (Novikoff et al. 1968; Stanka et al. 1981), NADH-ferricyanide oxidoreductase (Alroy et al. 1982; Morré et al. 1978), as well as acyltransferases (Higgins and Barrnett 1971) and adenylate cyclase (Cheng and Farquhar 1976) have been demonstrated by cytochemical methods in elements of the Golgi apparatus. Trimetaphosphatase (Doty et al. 1977), which has been suggested to label a subclass of lysosomes (Oliver 1980, 1983), appears in some cell types in elements of the Golgi complex as well; for example, all the stacked Golgi saccules in the lamina propria macrophages of the small intestine stain intensely for trimetaphosphatase (Fig. 14b), whereas in the absorptive enterocytes this enzyme is restricted to lysosomes (Fig. 14a).

By means of coupling uridine-5′-diphosphate formation and nicotinamide adenine dinucleotide reduction, galactosyltransferase has been demonstrated cytochemically in cisternae of Golgi apparatus isolated from rat liver (Matyas and Morré 1983).

Some of the methods listed above have been frequently employed for the labeling of distinct segments of the Golgi stacks; this is particularly true for the osmification techniques and the enzyme-cytochemical localizations of NADPase, TPPase, NDPase, and AcPase. However, the labeling patterns obtained by these techniques vary: (1) in several cell types, the respective reactions are limited to certain clear-cut subsections of the Golgi stacks, e.g., to cis, medial, trans, or transmost cisternae, whereas (2) in others, the labeling is distributed across all the Golgi subsections.

1. Differentiated staining of the Golgi subsections, such as is obtained in the mature pancreatic acinar cells: osmium label after prolonged osmification stains transport vesicles and saccules of the cis side of the Golgi stacks (Fig. 5a). TPPase (Fig. 5b) and NDPase (Fig. 5d = inosine diphosphatase, IDPase) are localized enzyme-cytochemically in one to three cisternae of the trans Golgi side and are reactive in coated structures and secondary lysosomes. The transmost Golgi cisternae and condensing vacuoles, showing slight TPPase and IDPase reactions, stain intensely for AcPase (Fig. 5c); furthermore, AcPase is reactive in several uncoated and cytoplasmically coated vesicles and secondary lysosomes.

The differentiated staining of cis and trans Golgi subsections underlines the polar construction of the stacks, which is morphologically apparent. In the author's studies with pancreatic acinar cells, the individual Golgi cisternae

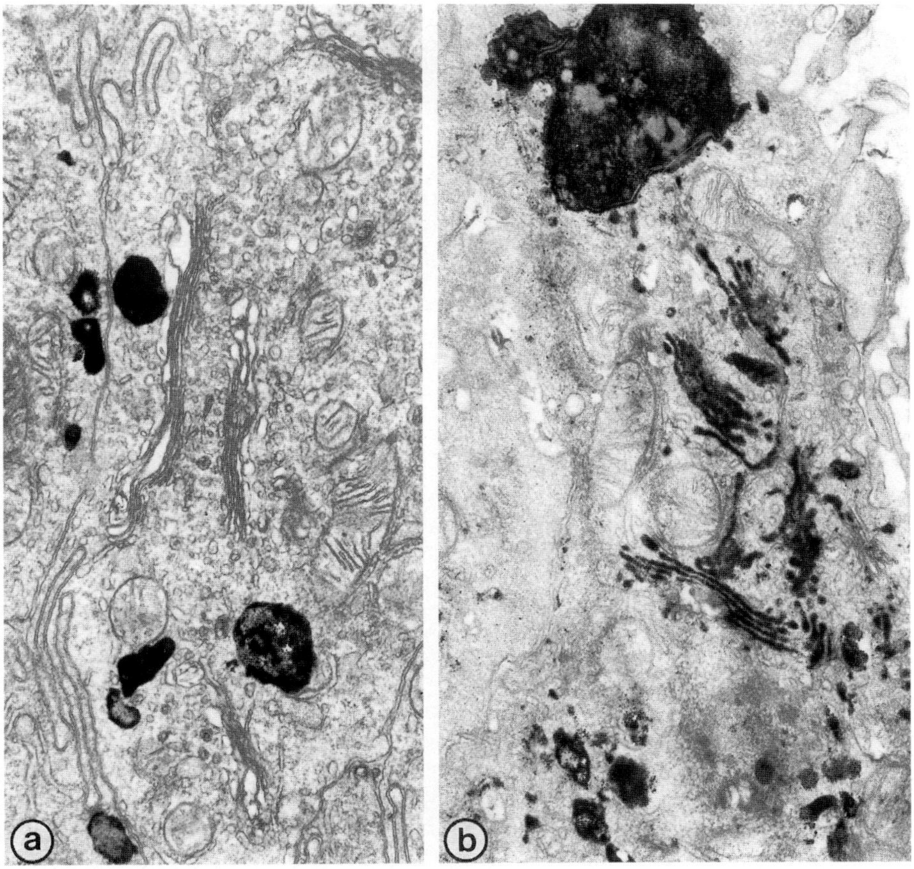

Fig. 14a, b. TMPase. **a** Absorptive cell of the rat duodenum. TMPase reaction is restricted to secondary lysosomes. Elements of the Golgi apparatus are devoid of staining. ×20000. **b** Macrophage in the lamina propria of the rat duodenum. In addition to secondary lysosomes, TMPase reaction products label most of the stacked Golgi cisternae and vesicles of the Golgi area. ×25000

were occupied not exclusively by one of the respective reactions; TPPase, IDPase, and AcPase reactions overlapped. By contrast, in other kinds of cells, e.g., in atrial muscle cells, each of the diverse reactions has been attributed to a separate cisterna of the stacks (Rambourg et al. 1984): the cismost osmium-labeled saccule is followed by an unreactive saccule; NADPase is located in one medial saccule, followed by one TPPase- and one CMPase-reactive saccule. Examination of several-micron-thick sections has shown that each of the cisternae continues into tubules that form an intersaccular system connecting the individual stacks of cisternae.

The reactions described above correspond to the most common labeling patterns; however, the reactions may differ between differently specialized cells. For example, comparative studies of diverse liver cells have demonstrated that, in hepatocytes, NADPase is located in the medial and trans saccules of the Golgi stacks, whereas, in the Kupffer cells, intense NADPase reaction labels the transmost Golgi elements (Angermüller and Fahimi 1984; Smith et al. 1986).

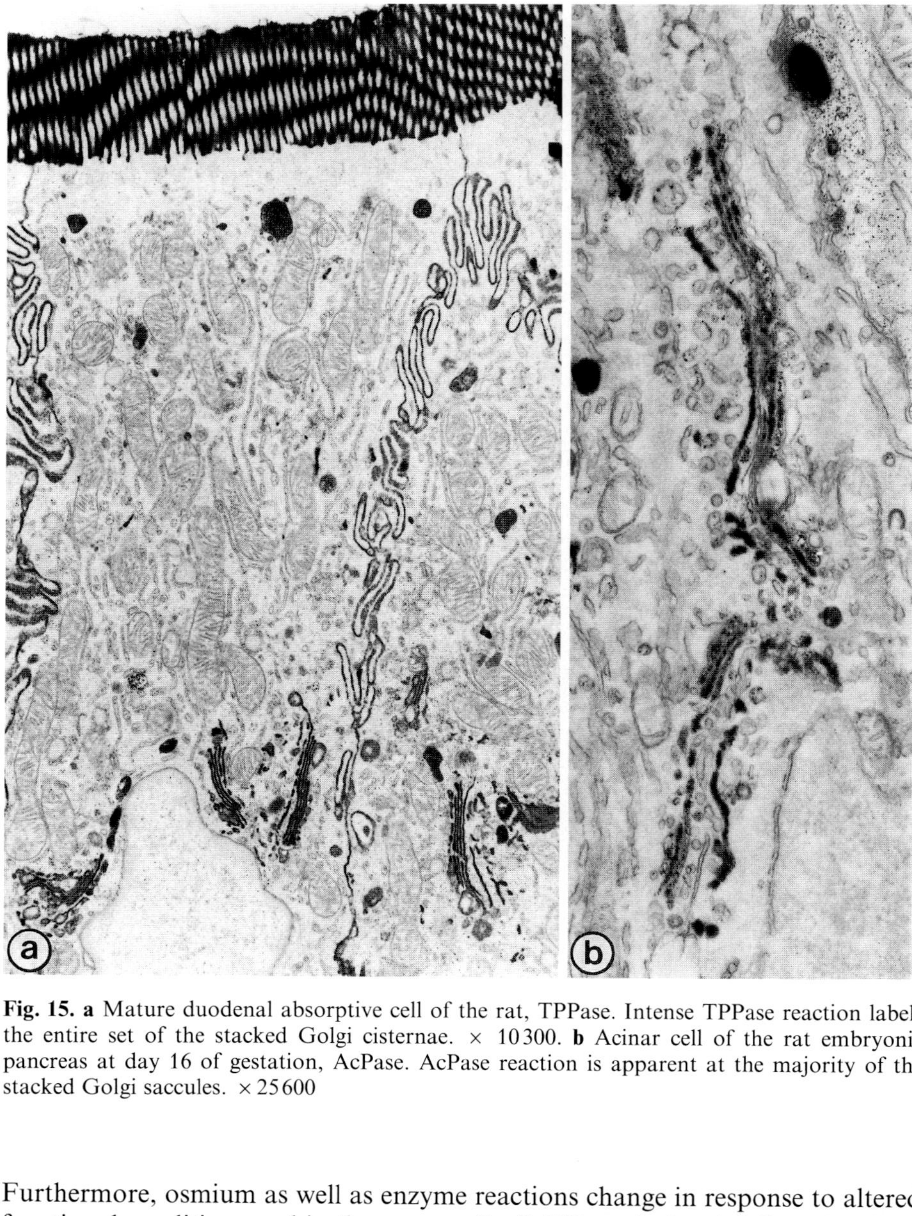

Fig. 15. a Mature duodenal absorptive cell of the rat, TPPase. Intense TPPase reaction labels the entire set of the stacked Golgi cisternae. × 10300. **b** Acinar cell of the rat embryonic pancreas at day 16 of gestation, AcPase. AcPase reaction is apparent at the majority of the stacked Golgi saccules. × 25600

Furthermore, osmium as well as enzyme reactions change in response to altered functional conditions and in the course of cell differentiation (cf. Sect. 4.1.1).

A special pattern is found in the mature absorptive cells of the jejunum (Pavelka and Ellinger 1981 a); here, TPPase and AcPase, such as cytochemical reactions for demonstrating complex carbohydrates (Sage and Jersild 1971), are localized in cisternae at the opposite side of those giving rise to lipid-filled secretory vesicles. In this case, both sides exhibit characteristics of the trans Golgi section.

2. In some cell types, and in connection with certain functional conditions, osmium as well as enzyme reactions are distributed in cisternae of each: cis, medial, trans, and transmost Golgi subsections (e.g., Bainton and Farquhar

1968; Beaudoin et al. 1983; Boutry and Novikoff 1975; Burchanowski et al. 1982; Friend 1969; Hickey et al. 1983; Kayahara 1982; Miller et al. 1980; Ono 1979; Oomori et al. 1984; Paavola 1978a, b; Quatacker 1979; Sasaki 1983; Wakayama et al. 1983/1984). The author's studies at absorptive cells of the rat duodenum (Pavelka and Ellinger 1982) showed intense TPPase and IDPase reactions in all the cisternae of the Golgi stacks, including the transmost one (Fig. 15a); similarly, in these cells, AcPase was distributed across the majority of the stacked Golgi saccules.

4.1.1 Enzyme Modulations

1. Changes of the enzyme-cytochemical patterns occur as a consequence of altered functional conditions of the cells (e.g., Broadwell and Oliver 1981; Decker 1974; Doine et al. 1984; Hand and Oliver 1984b; Oliver and Hand 1983; Oliver et al. 1980; Paavola 1978a, b; Smith and Farquhar 1970). Studies of secretory cells have shown that various conditions of increased secretory granule production are accompanied by enzyme redistribution (Hand and Oliver 1984b; Oliver and Hand 1983); during maximal secretory granule production, TPPase is not confined to the trans cisternae of the stacks, but extends to the transmost cisternae/rigid lamellae (GERL-like cisternae) and to condensing vacuoles, whereas, at the same time, AcPase frequently appears diminished.

2. Enzyme modulations also accompany cell differentiation (e.g., Doine et al. 1984; Hickey et al. 1983). The author's studies showed enzyme variations during the differentiation of exocrine pancreatic cells (Pavelka and Ellinger 1986a) and small intestinal absorptive cells (Ellinger and Pavelka 1982): in the protodifferentiated state of the pancreatic acinar cells, AcPase was restricted to the rigid lamellae system and small Golgi-associated vesicles. In the early differentiated state, most of the stacked cisternae reacted for AcPase (Fig. 15b), whereas the enzyme was again limited to the trans side in the late period of gestation. In the duodenum, the undifferentiated crypt cells showed AcPase as well as TPPase and IDPase limited to trans cisternae of the stacks. In the mature absorptive cells, located along the villi, the complete set of the stacked cisternae was intensely labeled for TPPase and IDPase, and the majority of the saccules reacted for AcPase (Fig. 15a; cf. this page, first paragraph).

The differentiated staining as obtained in a variety of cell types has led to the assumption that osmium and diverse enzyme reactions may be used as markers for characterizing subsections of the Golgi stacks: osmium for the cis side, NADPase for medial cisternae, TPPase for trans cisternae, and AcPase for the transmost Golgi elements. However, the variable patterns apparent in different kinds of cells and the enzyme modulations accompanying changes of the cell function and cell differentiation make it evident that Golgi architecture cannot be judged according to a fixed scheme; neither osmium nor the diverse enzyme reactions can be considered reliable markers for characterizing subsections of the Golgi stacks.

4.2 Immunocytochemistry

By means of immunocytochemistry, a wide variety of proteins has been localized in elements of the Golgi apparatus; they include newly synthesized molecules traversing the Golgi complex on their way to other cellular locations, such as secretory molecules, lysosomal enzymes, and plasma membrane constituents, as well as endogenous Golgi membrane constituents and internalized molecules.

4.2.1 Newly Synthesized Molecules Traversing the Golgi Apparatus

4.2.1.1 Secretory Molecules

Secretory constituents have been demonstrated immunocytochemically in all the Golgi area compartments of the secretory pathway, i.e., transitional elements of the endoplasmic reticulum, transport vesicles, cisternae of the Golgi stacks, and condensing vacuoles/secretory granules (e.g., Bendayan 1984; Bendayan et al. 1980; Brands et al. 1983; Geuze et al. 1979; Kraehenbuhl et al. 1977; Laurie et al. 1982a, b; Orci et al. 1984a; Posthuma et al. 1984; Ravazzola et al. 1981; Slot and Geuze 1983; Vassy et al. 1984; Wright and Leblond 1981). In contrast, rigid lamellae have, in most instances, been devoid of immunocytochemical reaction for secretory constituents (Bendayan 1984; Broadwell et al. 1979; Geuze et al. 1979; Hand and Oliver 1977a, b, 1981; Orci et al. 1984a).

Most secretory molecules, studied by immunocytochemistry, have been detected in all the Golgi cisternae, which indicates that the entire battery of the stacked saccules is involved in the secretory pathway. In the diverse Golgi subsections, different label intensities have been found, e.g., immunocytochemical staining for type IV collagen and laminin (Laurie et al. 1982a, b), although apparent in all the cisternae, has been shown to be most intense in the transmost saccules. Procollagen I has been found concentrated in the dilated peripheral segments of the saccules, the central plates being weakly reactive (Karim et al. 1979; Wright and Leblond 1981; cf. also Cho and Garant 1981a). In hepatocytes, immunoreaction for albumin has been found limited to some cisternae of the stacks, others being devoid of reaction (Feldmann et al. 1985; Guillouzo et al. 1982; Yokota and Fahimi 1981).

By using particulate markers, label densities have been determined (Bendayan 1984; Bendayan et al. 1980; Bergmann and Singer 1983; Geuze and Slot 1980; Posthuma et al. 1984). From the endoplasmic reticulum level to the level of the stacked Golgi cisternae, a steep increase has been ascertained, which suggests that condensation processes may occur already at the cis side of the stacks (Bergmann and Singer 1983; Posthuma et al. 1984; Slot and Geuze 1983). It has been shown that the degrees of concentration are different for diverse secretory molecules, e.g., in pancreatic acinar cells, being higher for chymotrypsinogen as opposed to amylase (Geuze and Slot 1980; Posthuma et al. 1984). Furthermore, at the level of the Golgi stacks, concentration of secretory materials has been found to occur predominantly at the trans side (Bendayan 1984).

Colocalization

Various secretory products, such as albumin, transferrin, and lipid particles, have been demonstrated immunocytochemically in identical cisternae and vesicles (Strous et al. 1983b). Moreover, secretory molecules colocalize with lysosomal enzymes and membrane constituents traversing the Golgi apparatus (Geuze et al. 1984a, 1985; Slot and Geuze 1983). In hepatocytes, albumin and lipid particles have been observed in separate cisternae of the cis side of the Golgi stacks, although colocalizing in cisternae of the trans side and in secretory vesicles (Yokota and Fahimi 1981).

4.2.1.2 *Lysosomal Enzymes*

The distribution in the Golgi apparatus of fibroblasts and hepatoma cells of a series of lysosomal enzymes has been studied (Geuze et al. 1985; van Dongen et al. 1984); each of the enzymes investigated has been detected in the entire set of the stacked cisternae. In contrast, another investigation found immunoreactivity for β-galactosidase in hepatocytes to be limited to elements of the trans Golgi side referred to as GERL (Novikoff et al. 1983a).

4.2.1.3 *Membrane Proteins*

Various membrane proteins traversing the Golgi apparatus on their way to other cellular destinations have been immunocytochemically localized in elements of the stacks (e.g., Danielsen et al. 1986; Fransen et al. 1985): the labeling either has been found uniformly distributed across the stacks (Fransen et al. 1985) or predominating on the trans side (Danielsen et al. 1986). Furthermore, immunostaining of the Golgi apparatus has been obtained for a number of virus glycoproteins, being widely used as models for studying intracellular transport of membrane proteins (e.g., Bergmann and Singer 1983; Bergmann et al. 1981; Burke et al. 1983; Green et al. 1980; Griffiths et al. 1983; Rindler et al. 1985; Saraste and Kuismanen 1984; Strous et al. 1983b); immunoreactivity across the entire stacks has been shown for most molecules investigated. Studies with doubly infected cells have revealed that viral glycoproteins destined for either apical or basolateral plasma membrane domains traverse the same Golgi stacks (Fuller et al. 1985; Rindler et al. 1984a). They apparently are not sorted until they are exported from the stacks of Golgi cisternae. It is the compartment of trans-located, Golgi-associated elements that is considered as being crucial for the sorting processes (Rindler 1986).

Various Membrane Receptors

The receptor of asialoglycoproteins has been localized in the Golgi apparatus of hepatocytes (Geuze et al. 1982; Mizuno et al. 1984); immunolabeling has been found in the entire set of the stacked cisternae, being strongest at the trans side (Slot and Geuze 1983).

In contrast to the reaction in the stacks of the asialoglycoprotein receptor, in KB carcinoma cells, immunostaining for the receptors of the epidermal growth factor (Beguinot et al. 1984) and transferrin (Willingham and Pastan 1985a) has been shown to predominate in the tubular reticulum located in the transmost position; in the trans cisternae of the stacks, only a small amount of the transferrin receptor has been detected. The location in the transmost tubular elements of the epidermal growth factor and transferrin receptors corresponds to the location of the respective internalized ligands (Hanover et al. 1984; Willingham and Pastan 1982; Willingham et al. 1983a, 1984).

Golgi localization of the intrinsic factor-cobalamine receptor has been shown in the ileal epithelial cells (Levine et al. 1984).

The immunocytochemical results as regards the 215-kDa mannose-6-phosphate receptor, being one of the crucial determinants for targeting of newly synthesized lysosomal enzymes from the Golgi apparatus to lysosomes (Creek and Sly 1984; Sly and Fischer 1982), are at variance: in hepatocytes, Geuze and co-workers (Geuze et al. 1984a) have localized the phosphomannosyl receptor across the entire stacks of Golgi cisternae, being concentrated in the lateral outwrappings of the cisternae. In contrast, studies of Brown and Farquhar (Brown and Farquhar 1984a), with a number of different kinds of cells, have revealed immunoreactivity for the 215-kDa mannose-6-phosphate receptor to be limited to cisternae of the cis Golgi side; Willingham and co-workers have shown trans localization of phosphomannosyl receptor in Chinese hamster ovary cells (Willingham et al. 1983b).

A second mannose-6-phosphate receptor, being distinct from the 215-kDa protein and requiring divalent cations for ligand binding, has recently been characterized (Hoflack and Kornfeld 1985).

For the phosphomannosyl receptor (Geuze et al. 1984c), such as for the asialoglycoprotein receptor (van den Bosch et al. 1986), it has been shown that a cycloheximide-resistant pool exists in the Golgi complex of hepatocytes, which indicates receptor recycling.

4.2.2 Enzymes Involved in Glycan Biosynthesis

N-acetyl-glucosaminyltransferase I, initiating formation of complex-type asparagine-linked oligosaccharides, has been localized immunocytochemically in cisternae of the medial subsection of the Golgi stacks (Dunphy et al. 1985).

Galactosyltransferase was the first sugar transferase to be demonstrated by immunocytochemistry in the Golgi apparatus (Roth and Berger 1982; Strous et al. 1983a). In HeLa cells, immunocytochemical galactosyltransferase reactions have been shown to be limited to the trans Golgi side, including the trans cisternae of the stacks as well as tubular-reticular elements in the transmost position. Immunofluorescence studies of kidney cells and fibroblasts have revealed the intracellular locations of galactosyltransferase to be different from those of sialyltransferase (Berger and Hesford 1985): cellular structures stained for galactosyltransferase have been found in juxtanuclear position and crescent-shaped. Sialyltransferase immunoreaction has been localized in cytoplasmic vesicles concentrated around an unstained Golgi area or spread over the whole cytoplasm.

By contrast, studies with monensin-treated HeLa cells indicate that galactosyltransferase is located in the same compartments in which sialyltransferases are active (Strous et al. 1985b). Sialyltransferase has been demonstrated at an extensive interconnected cisternal-tubular system at the trans Golgi side of hepatocytes, which consists of the trans cisternae of the stacks and a tubular network in the transmost position (Roth et al. 1985). By contrast, in the absorptive cells of the colon, a sialyltransferase as well as blood group A α 1,3 N-acetylgalactosaminyltransferase have been found distributed across all the stacked Golgi saccules, except for the cismost one (Roth et al. 1986). This variability is consistant with the variable distribution in the Golgi apparatus of binding sites for the sialic acid-recognizing LFA (Hedman et al. 1986; Roth et al. 1986).

Golgi mannosidase II, being active in the trimming of α-1,6 mannosyl residues in connection with the synthesis of complex-type oligosaccharides, has been localized immunocytochemically in all cisternae of the Golgi stacks of rat liver cells (Novikoff et al. 1983b).

4.2.3 Antibodies Directed Against Golgi-Associated Molecules

With the aid of (monoclonal) antibodies raised against Golgi-associated proteins (Burke et al. 1982; Lin and Queally 1982; Louvard et al. 1982; Smith et al. 1984; Tougard et al. 1983), distinct, biochemically well-defined Golgi subcompartments have been visualized; in each of the preparations, the reactions have been localized in clear-cut subregions of the stacks.

These results include the immunolocalization of two cytoplasmically oriented membrane proteins considered as being involved in membrane-membrane interactions for maintaining the stacked organization of the Golgi cisternae (Chicheportiche et al. 1984).

Furthermore, antibodies directed against constituents of the Golgi apparatus have been found in the serum of patients suffering from autoimmune diseases (Fritzler et al. 1984; Rodriguez et al. 1982).

4.2.4 Lysosomal Membrane Antigens

It has been shown that antibodies raised against a lysosomal membrane antigen, presumably representing a proton pump component, recognize elements of the Golgi apparatus (Reggio et al. 1984). Several cell types, including hepatocytes, kidney cells, macrophages, and prolactin cells, have been examined, showing cis and/or medial saccules of the Golgi stacks to be reactive (Tougard et al. 1985). In contrast, no reaction has been detected in the transmost Golgi elements. The immunocytochemical findings harmonize with recent biochemical results concerning a Golgi membrane-associated electrogenetic pump, being responsible for acidification of Golgi-derived vesicles (Barr et al. 1984; Glickman et al. 1983; Zang and Schneider 1983).

Recently, two other lysosome-associated membrane proteins have been localized immunocytochemically in trans elements of the Golgi stacks (Chen et al. 1985).

4.2.5 c-AMP-Dependent Protein Kinase Type II

By immunocytochemistry, regulatory and catalytic subunits of c-AMP-dependent protein kinase type II have been found concentrated in the Golgi area as well as associated with microtubules and microtubule-organizing centers (DeCamilli et al. 1986; Nigg et al. 1985a). In the case of regulatory subunits, immunoreactions have been localized at the trans side of the Golgi stacks (DeCamilli et al. 1986). Localization in the Golgi area of protein kinase II may reflect importance in mediating effects of cAMP on processes associated with the Golgi apparatus: control of intracellular membrane transport has been considered as well as influence on the organization of cytoskeletal elements, the latter possibly being connected with phosphorylation of microtubule-associated protein II (DeCamilli et al. 1986; Nigg et al. 1985a). The list of further potential phospho-acceptor molecules may include proteins traversing the Golgi apparatus as well as endogenous Golgi molecules. Furthermore, it has been shown that the catalytic, but not the regulatory, subunit of protein kinase II rapidly translocates from the Golgi area to the nucleus upon activation of cAMP by treatment of cells with the adenylate cyclase activator forskolin (Nigg et al. 1985b).

4.3 Lectin Cytochemistry

Lectins are proteins of plant or animal origin capable of binding specifically with certain sugars or sugar sequences (Brown and Hunt 1978; Debray et al. 1981; Goldstein and Hayes 1978; Irimura and Nicolson 1983; Sharon and Lis 1972). At the light microscopic level, labeled lectins have widely been employed as histochemical probes for investigating cellular and extracellular glycoconjugates. Electron microscopic lectin binding reactions make it possible to locate intracellularly glycoconjugates containing saccharide chains of certain sugar sequences. This is particularly interesting as regards the elements of the Golgi apparatus being the sites of insertion of sugar residues into glycans. Lectin cytochemistry at the electron microscopic level may be considered a hopeful approach for labeling Golgi subsections involved in the step-by-step addition of sugar molecules in the growing saccharide chains and, hence, for distinguishing functionally different Golgi subcompartments.

In the wide range of lectins, several have revealed binding reactions that are particularly valuable in the interpretation of Golgi architecture: concanavalin A (ConA), which preferably binds with α-D-mannosyl and α-D-glucosyl residues; *Ricinus communis I* and *II* agglutinins (RCA I/II), which recognize β-D-galactose and β-D-galactose/N-acetyl-galactosamine, respectively; peanut lectin (PNA), which reacts with N-acetyl-galactosamine-galactose sequences; wheat germ agglutinin (WGA), binding with N-acetyl-glucosamine/sialic acid; *Limax flavus* lectin (LFA), which is specific for sialic acid; *Helix pomatia* agglutinin (HPA), which is a N-acetyl-galactosamine probe; and *Pisum sativum* as well as *Lens culinaris* lectins (PSA, LCA), which bind with mannose/glucose residues, a fucose residue attached to the asparagine-linked N-acetyl-glucosamine being essential for high-affinity binding to glycopeptides (Debray et al. 1981; Kornfeld et al. 1981).

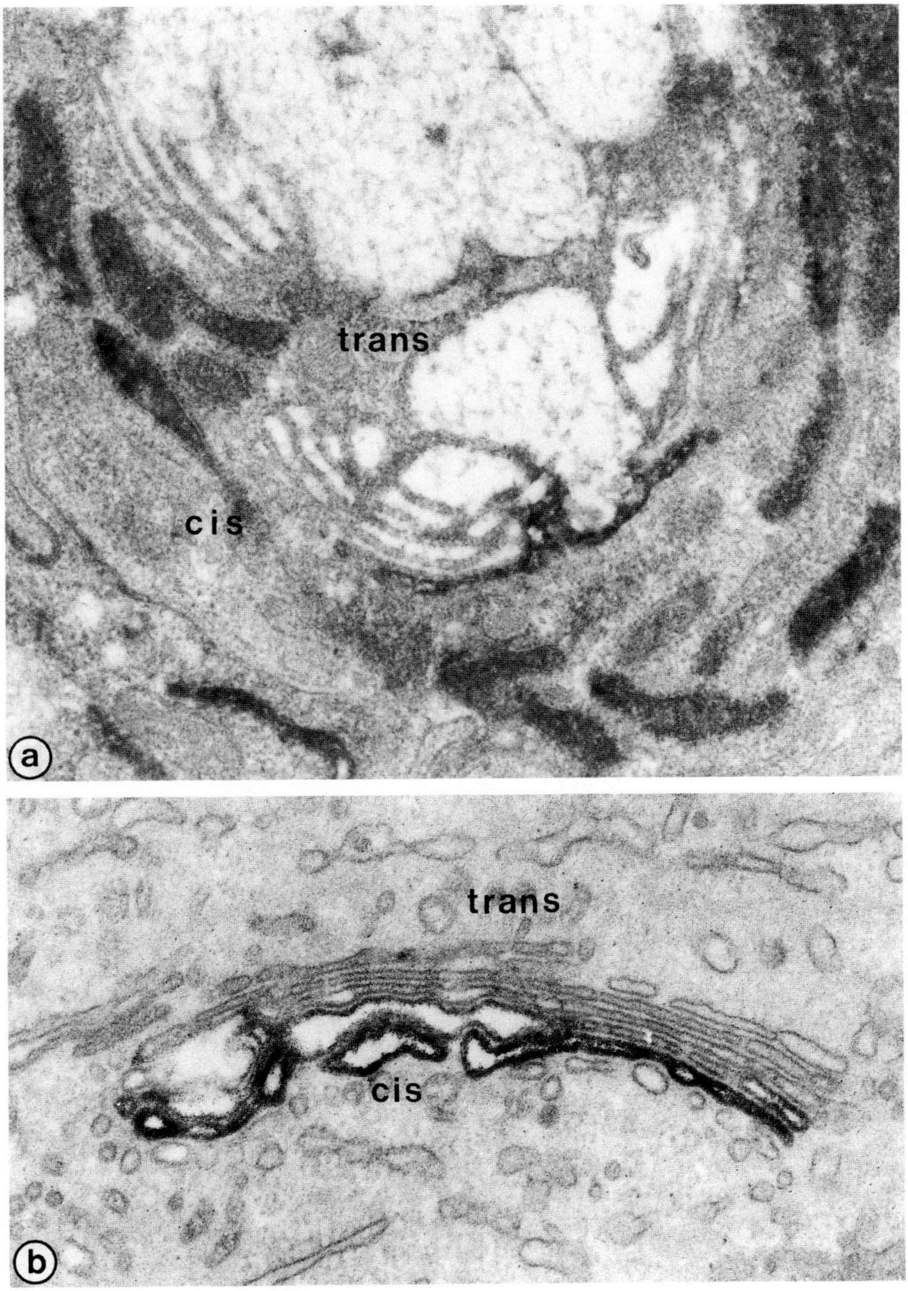

Fig. 16a, b. ConA. ConA binding reactions, visualized by the two-step ConA-HRP method. **a** Rat duodenum, goblet cell. ConA reactions label cisternae of the endoplasmic reticulum as well as one cis Golgi saccule. ×45000. **b** Rat duodenum, absorptive cells. ConA label is confined to two cisternae of the cis section of this stack. ×50000

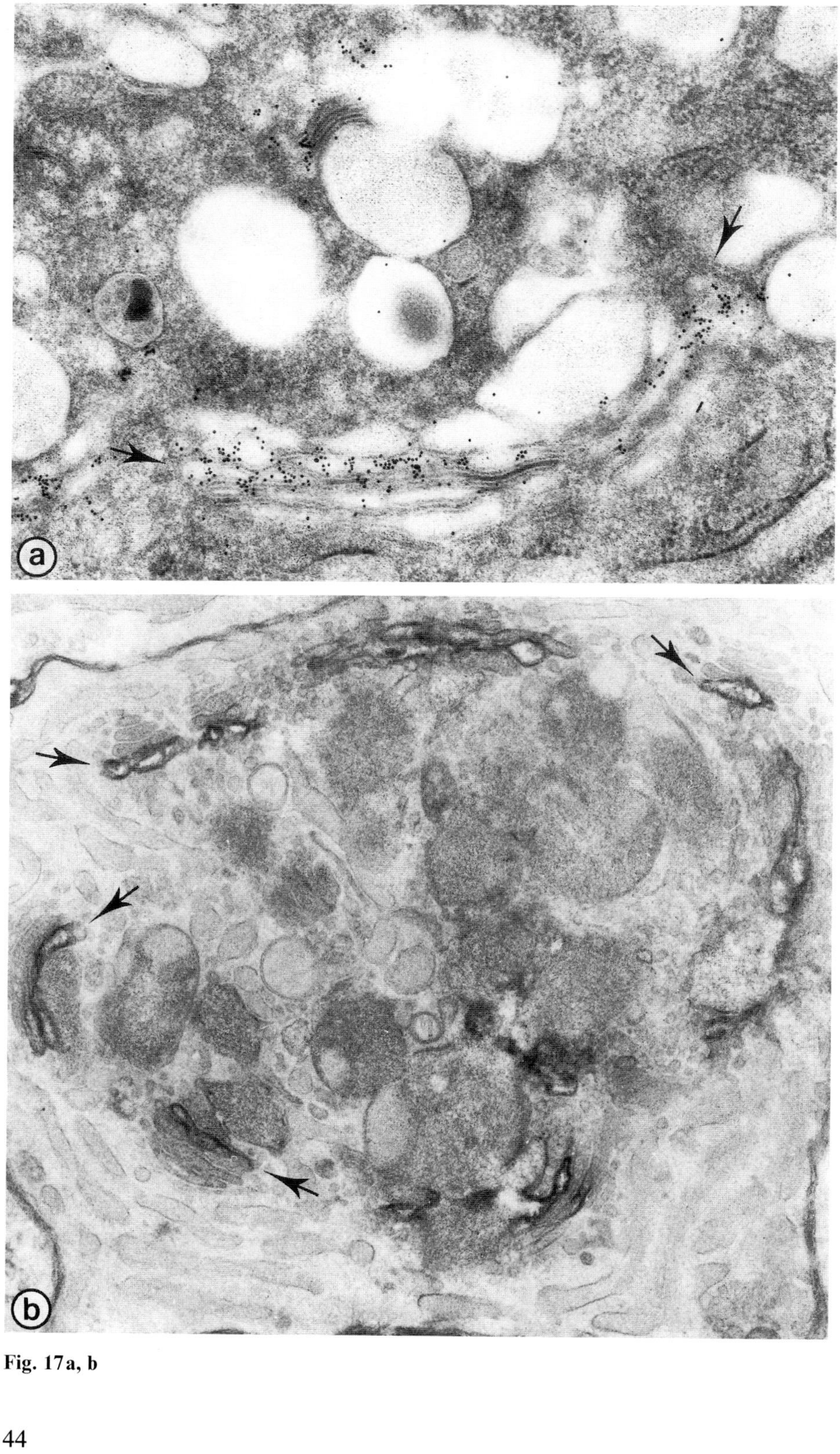

Fig. 17a, b

The majority of the lectin binding patterns known so far show differentiated labeling of certain distinct regions of the Golgi apparatus; these may be subsections of the stacks, e.g., cis, medial, trans, or transmost cisternae, or subregions of the individual cisternae, such as peripheral extremities or central plates. Binding sites for ConA are preferably located in cisternae of the cis Golgi side (Pavelka and Ellinger 1985; Tartakoff and Vassalli 1983; Figs. 16, 18a). RCA I preferentially binds with constituents of medial and trans cisternae (Griffiths et al. 1982; Pavelka and Ellinger 1985, 1986c; Roth 1983; Figs. 17, 18b); reactions for RCA II are strongest in trans Golgi elements (Gorelick et al. 1982). PNA reacts with constituents of medial cisternae (Sato and Spicer 1982a, b); WGA labels medial/trans cisternae (Tartakoff and Vassalli 1983); LFA, in hepatocytes, binds with tubular reticular elements located in the transmost position of the Golgi stacks (Roth et al. 1984). By contrast, studies with 3T3 fibroblasts (Hedman et al. 1986) and absorptive cells of the colon (Roth et al. 1986) showed LFA binding reactions in most of the stacked cisternae, only one cisterna at one side being unlabeled. HPA binding sites are preferentially cis located (Pavelka and Ellinger 1985; Roth 1984; Fig. 21b, c); PSA and LCA favorably react with constituents of the cis/medial saccules (Figs. 22a, 23a).

So far, a limited number of cell types has been investigated; the author's studies particularly concentrated on the distribution in Golgi elements of binding sites for ConA, RCA I, HPA, PSA, and LCA.

4.3.1 Concanavalin A

ConA bound intracellularly with constituents of the endoplasmic reticulum, Golgi apparatus, and lysosomes. Cisternae of the cis side were the predominant locations of ConA binding sites in the Golgi apparatus. In the small intestinal goblet cells, ConA reactions were limited to cis cisternae (Fig. 16a); in the absorptive cells of the small intestine, cis localization of ConA binding sites was dominant (Fig. 16b). However, distribution of the ConA label across the entire battery of the stacked cisternae was observed in approximately 15% of the stacks (Pavelka and Ellinger 1985).

Considerable variability of the ConA reactions was found in the acinar cells of the embryonic pancreas; in each developmental stage, ConA binding sites were either confined to cis and/or medial cisternae (Fig. 18a) or were distributed in elements of cis, medial, and trans cisternae, including the transmost one. Unreactive saccules were interposed in between strongly reactive cis and trans cisternae (Fig. 20a). In the stacks with cis-confined reaction, the staining intensity was usually strongest in the cismost cisterna and decreased toward the medial section of the stacks (Fig. 18a). Within the individual cisternae, variable label intensities were found, intensely labeled segments alternating with faintly stained and unreactive portions (Fig. 18a).

◁ **Fig. 17a, b.** RCA I. RCA I binding reactions, visualized by colloidal gold label (**a**) and HRP (**b**), predominate in the medial and trans cisternae (→) of these Golgi stacks of goblet cells of the rat duodenum. **a** ×50000; **b** ×41000

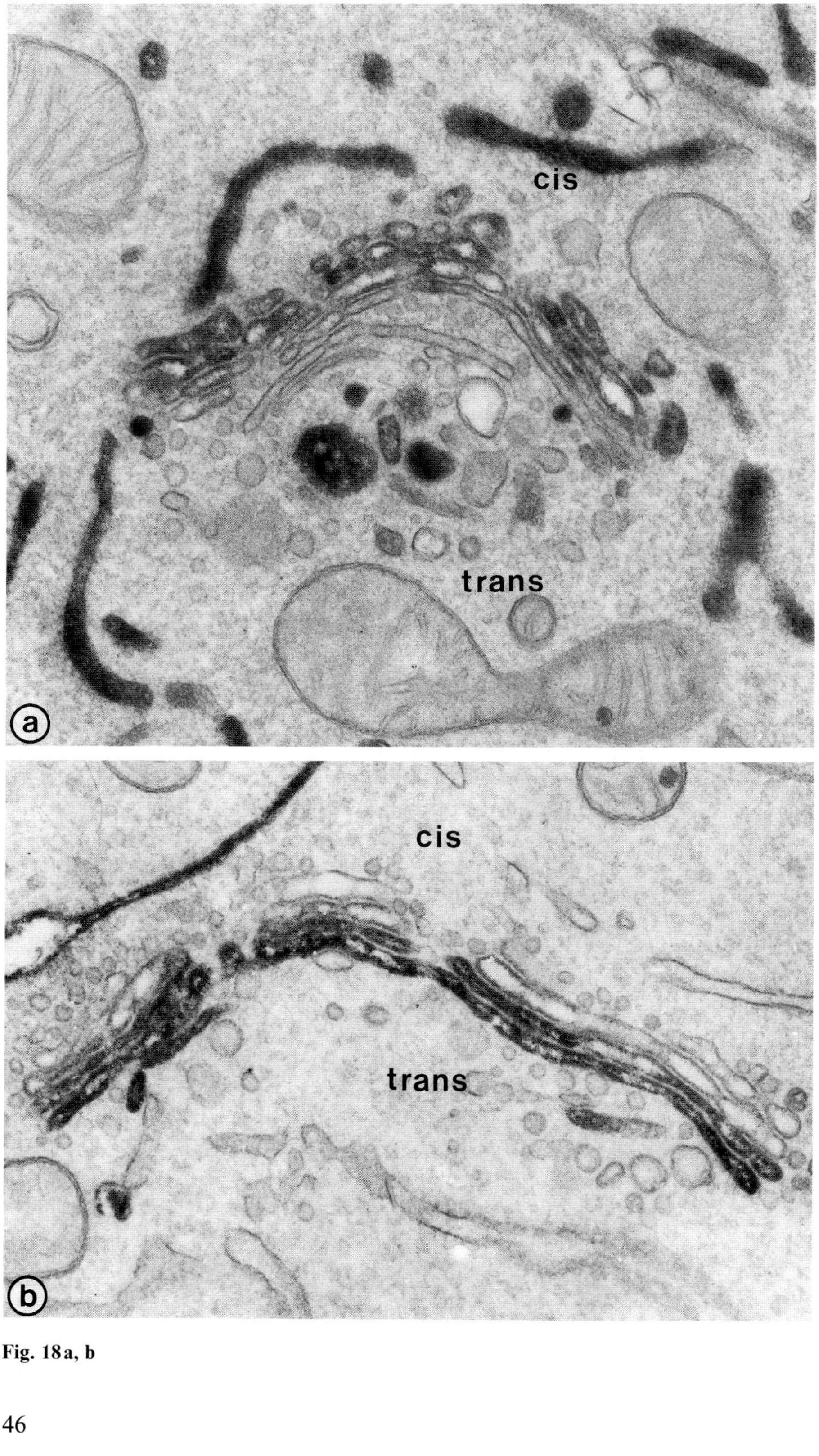

Fig. 18a, b

4.3.2 Ricinus communis I Agglutinin

Binding sites for RCA I were intracellularly localized in the Golgi apparatus, in several vesicles of the Golgi area, and in lysosomes. In the Golgi apparatus, intense label of the cisternae of the trans side and weak staining of medial cisternae was the dominant RCA I reaction pattern (Figs. 17a, b, 18b). In the small intestinal goblet cells, this pattern was consistently found (Fig. 17a, b). In the absorptive enterocytes, trans cisternae were stained, although RCA I binding sites were concentrated in the medial cisternae of the stacks. In the embryonic acinar cells, RCA I reactions labeled the extended rigid lamellae system apparent in early stages of differentiation (Figs. 6b, 20b). In the stacked cisternae, the reactions were variable; galactosyl residues accessible to RCA I were either restricted to medial/trans saccules (Figs. 6b, 18b) or were found distributed in cisternae of each subsection of the stacks (Figs. 19a–c, 20b).

As regards the respective labeling patterns, similar percentages were found for the early and late stages of development, viz., ~62% for the reaction restricted to medial/trans cisternae and ~38% for the "cis-to-trans-distributed" reaction (Pavelka and Ellinger 1986c). In the "cis-to-trans"-reactive stacks, frequently the label was weak in the cis saccules and increased in intensity toward the trans side (Fig. 19). In other cases, unstained saccules were interposed in between intensely stained cisternae (Fig. 20b). Within the individual saccules, the reaction intensities varied (Figs. 19a, 20b). Furthermore, stacks with different reaction patterns were located side by side (Fig. 19c).

4.3.3 Helix pomatia Lectin

HPA reactions were localized intracellularly in the Golgi area. In goblet and absorptive cells, intense label was found in cisternae of the cis side of the stacks (Fig. 21b, c); frequently, the reaction was most intense in the penultimate cisterna of the cis side, the cismost cisterna being faintly stained or unreactive (Fig. 21b). In duodenal goblet cells of blood group A-positive animals, HPA reactions have been localized in cis as well as trans saccules, the medial cisternae being devoid of label (Roth 1984). In embryonic acinar cells, HPA binding reactions were, in each differentiation stage, either limited to the cis side or were found in cisternae of each Golgi subsection. Similarly, HPA label was found in all the stacked cisternae of the Golgi apparatus in serous cells of the submandibular gland (Fig. 21a); the reaction intensities varied considerably between different portions of the individual cisternae.

◁ **Fig. 18a, b.** ConA and RCA I. Acinar cells of the rat embryonic pancreas at day 16 of gestation. **a** ConA. ConA reactions label endoplasmic reticulum cisternae, saccules of the cis and medial Golgi subsections, vesicles of the Golgi area, and a multivesiculated body. The reactions are more intense in the cis as compared with the medial saccules; within the individual cisternae, the reaction intensities are variable. ConA-HRP two-step technique. ×42500. **b** RCA I. Reaction products indicating presence of RCA I binding sites are concentrated in medial and trans saccules of this stack. RCA I-HRP conjugates. ×42500

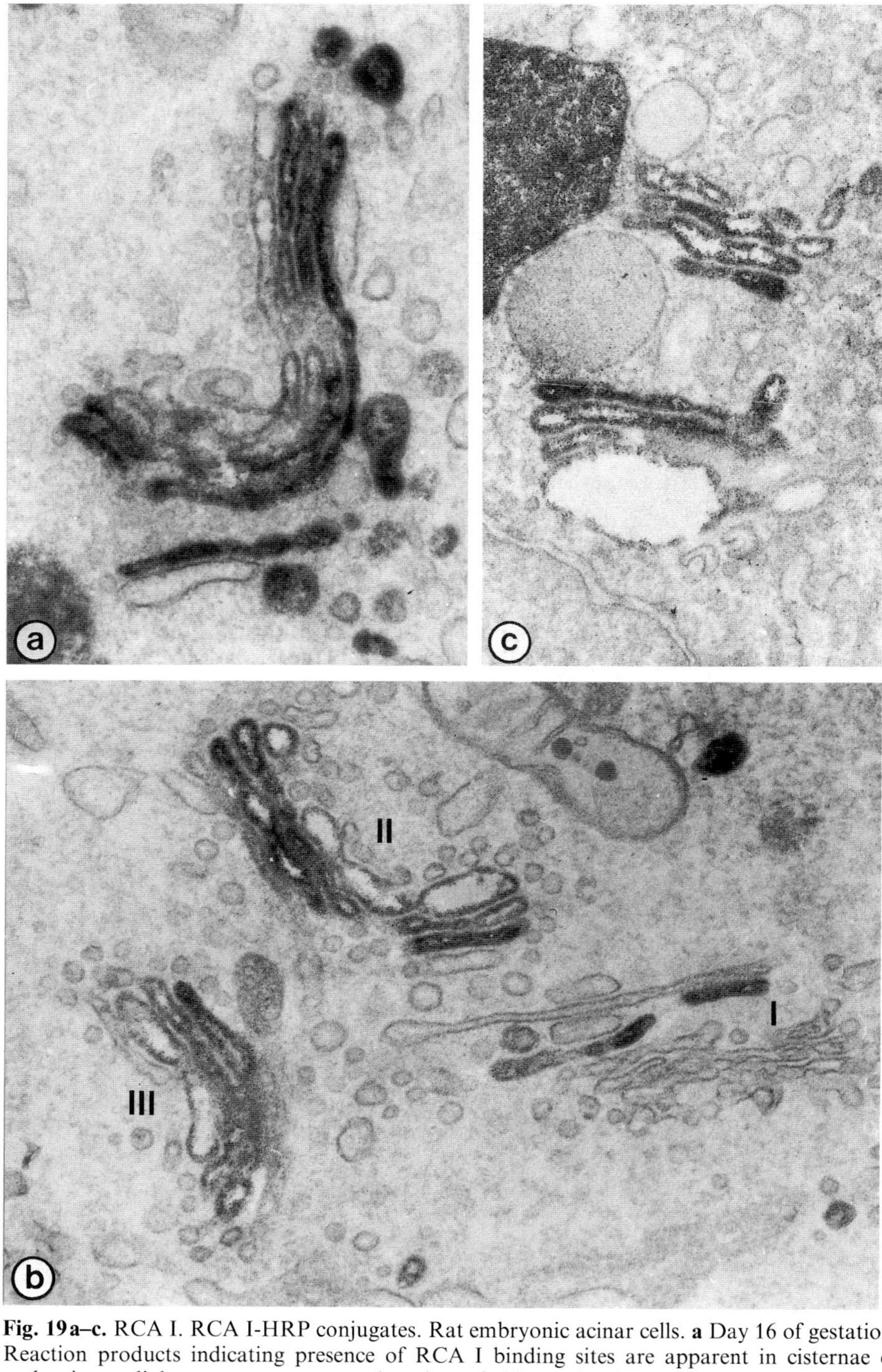

Fig. 19 a–c. RCA I. RCA I-HRP conjugates. Rat embryonic acinar cells. **a** Day 16 of gestation. Reaction products indicating presence of RCA I binding sites are apparent in cisternae of each, cis, medial, trans, transmost, subsection of this stack: the reactions are weak in the cis cisternae and increase in intensity toward the trans side. × 51 000. **b** Day 16 of gestation. Golgi stacks with diverse RCA I labeling patterns reside side by side: the reactions either are confined to trans cisternae (*I*) or apparent in most of the stacked saccules (*II* and *III*). × 50 000. **c** Day 19 of gestation. All the cisternae of these stacks exhibit RCA I binding reactions. × 48 000

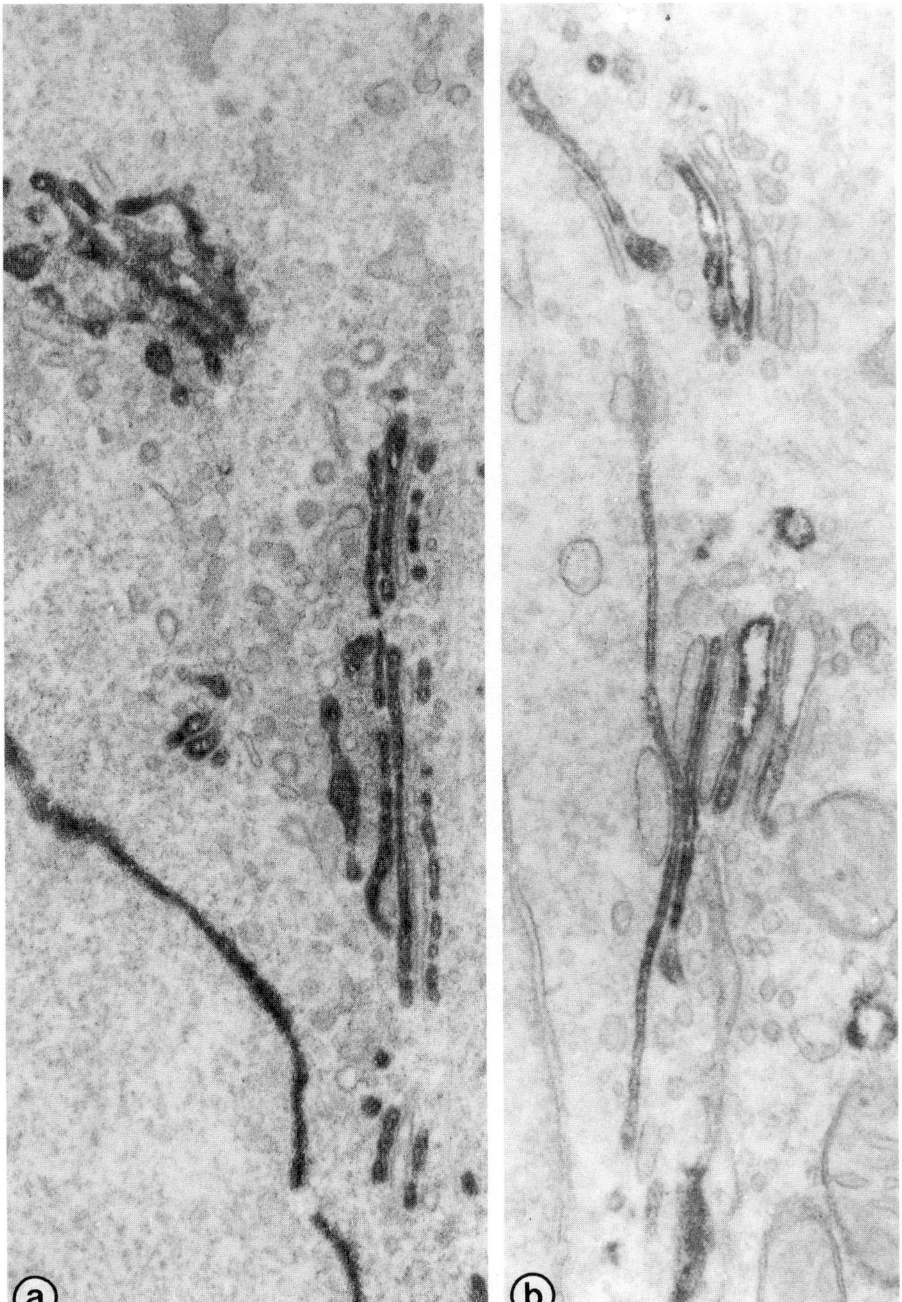

Fig. 20a, b. ConA and RCA I. Rat embryonic pancreatic acinar cells, day 16 of gestation. **a** ConA. Two-step ConA-HRP method. Intense reactions indicating presence of ConA binding sites are apparent in cisternae of the cis as well as trans and transmost Golgi subsections, an unstained medial saccule being interposed. ×41000. **b** RCA I. RCA I-HRP conjugate. RCA I binding reactions are apparent in one saccule of each of the cis, medial, trans, and transmost subsections of this stack; unstained cisternae are interposed between the reactive cis and medial, the medial and trans, as well as the trans and transmost saccules, respectively. ×44000

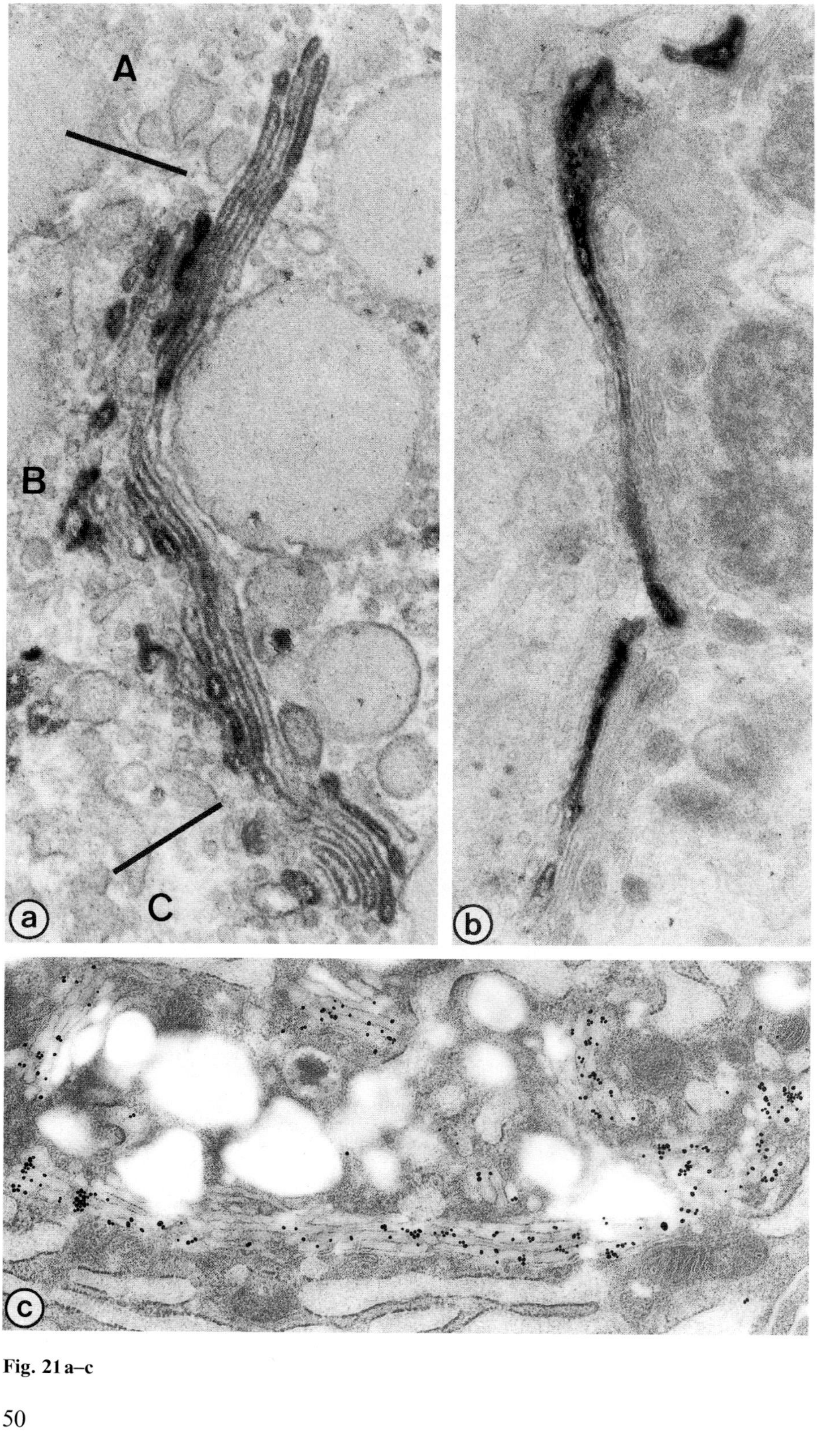

Fig. 21 a–c

4.3.4 Pisum sativum Lectin

In small intestinal absorptive cells, intense PSA reactions were intracellularly localized in the Golgi apparatus and in lysosomes; weak staining was found in some limited segments of the endoplasmic reticulum. In the Golgi apparatus, PSA binding caused intense reaction in cis and/or medial cisternae (Fig. 22a); frequently, staining was most intense in the penultimate cisterna at the cis side, the cismost saccule being weakly stained or lacking reaction product. In several stacks, PSA binding was found in limited segments of trans cisternae.

In the small intestine, two types of goblet cells were found. One showed PSA binding reactions of materials contained in the secretory granules; in these cells, the PSA Golgi reaction was weak in the cis cisternae and increased in intensity toward the trans side (Fig. 22b). In other goblet cells, the secretion granules lacked PSA reaction products; here, the PSA Golgi reaction was limited to the cis/medial cisternae, the penultimate cis cisterna being the favored one.

4.3.5 Lens culinaris Lectin

In the mature absorptive enterocytes, LCA caused weak and dotted reaction of the endoplasmic reticulum; intense reaction was found in the Golgi apparatus and in lysosomes. The Golgi reaction essentially corresponded to that obtained with PSA; the reaction was strong in cis and/or medial saccules (Fig. 23a).

◁ **Fig. 21a–c.** HPA. **a** Acinar cell of the rat submandibular gland. Mosaiclike staining. HPA-HRP conjugate. Within the individual cisternae, the intensities of the HPA reactions vary considerably, intensely stained portions of the cisternae alternating with faintly stained and unreactive portions. As a consequence, in the different segments of this stack, different HPA binding patterns are apparent: segments *A* and *C* show concentration of HPA binding sites in cisternae of one side, which presumably is the trans side; in segment *B*, the most intense HPA label is apparent in cisternae of the opposite (cis?) side. ×51000. **b** Goblet cell of the rat duodenum. HPA-HRP conjugate. HPA binding sites are concentrated in the penultimate cisterna of the cis side of this stack; trans and transmost cisternae as well as the cismost cisterna are devoid of label. ×51000. **c** Goblet cell of the rat duodenum. HPA-gold conjugate. Cis and medial cisternae of this stack are labeled. ×41000

Fig. 22a, b. PSA. PSA-HRP conjugates. **a** Rat duodenum, mature absorptive cell. Reactions indicating presence of PSA binding sites are concentrated in cisternae of the cis Golgi side, the penultimate cis cisterna being most intensely labeled (⟶); weak reaction is apparent in limited regions of trans saccules (➤). *Ly*, lysosome. ×32000. **b** Rat duodenum, goblet cell. PSA staining is weak in the cis cisternae and increases in intensity toward the trans side; trans cisternae as well as secretory granules are intensely labeled. ×30000

Fig. 23a, b. LCA. LCA-HRP conjugates. **a** Rat duodenum, mature absorptive cell. The cismost cisterna is weakly stained; LCA reaction products are concentrated in the penultimate cisterna of the cis side (⟶) and in one saccule of the transmost subsection of this stack (➤). ×51000. **b** Rat duodenum, epithelial cell of the crypt-top region. Mosaiclike staining. LCA reactions are distributed across the majority of the stacked Golgi cisternae, although within the individual cisternae the reaction intensities are variable. This mosaiclike staining accounts for the appearance of different LCA binding patterns in the different segments of the stack: segments *A* exhibit LCA label limited to one side of the stack; by contrast, in segments *B*, LCA reaction is extended across most or the entire set of the stacked cisternae. ×51000

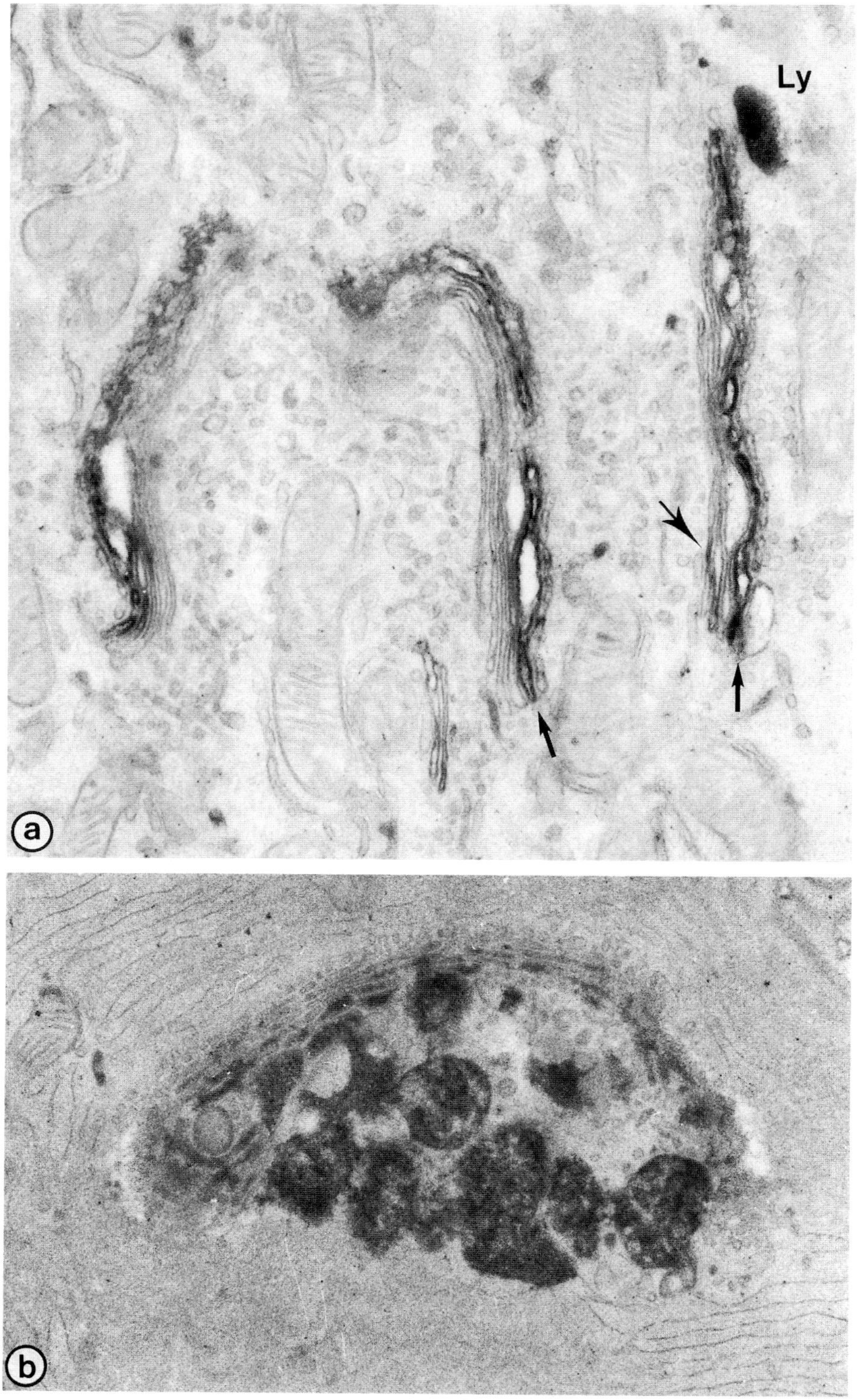

Fig. 22a, b. Legend see p. 51

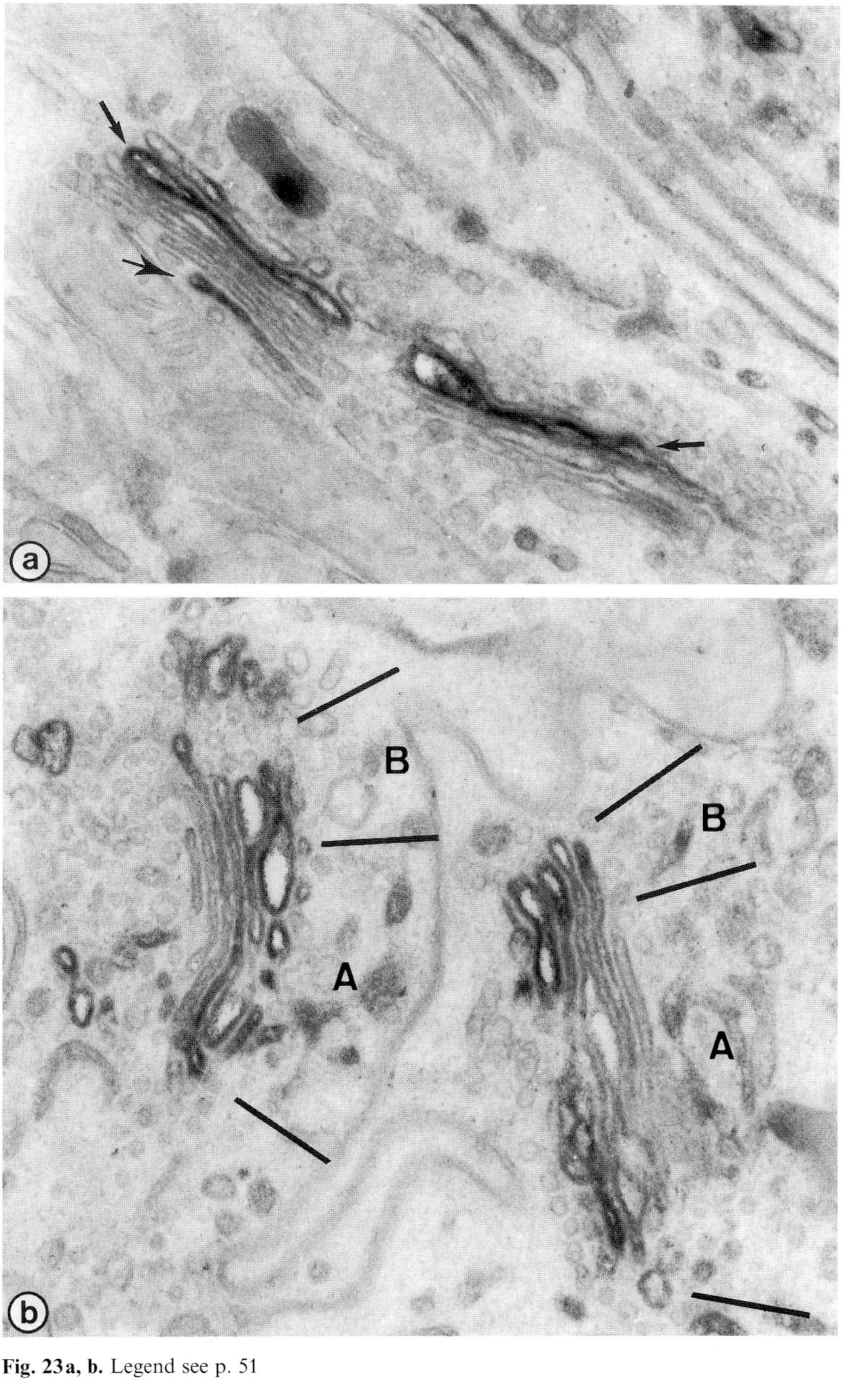

Fig. 23a, b. Legend see p. 51

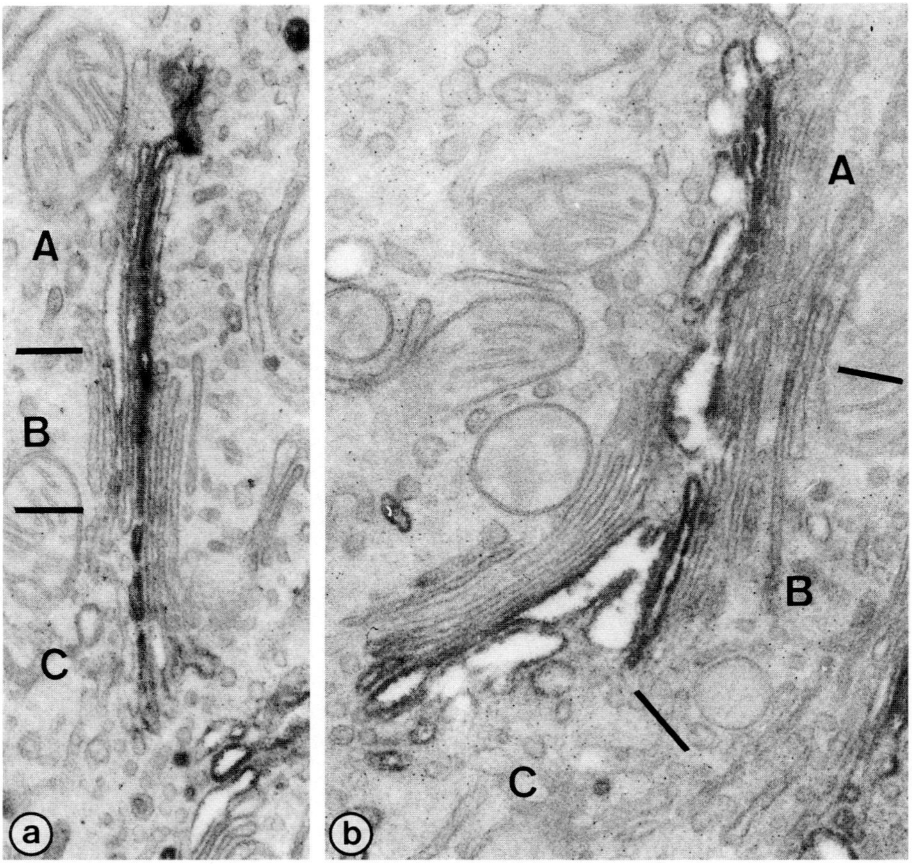

Fig. 24a, b. PSA. PSA-HRP conjugates. **a, b** Rat duodenal absorptive cells, backbone cisternae. Cisternae labeled by PSA reaction products constitute diverse positions within the stacks: in segments *A* of the stacks, PSA-reactive cisternae are in the ultimate (**b**) or penultimate (**a**) cis position; in segments *B*, the same PSA-positive cisternae are located in the medial subsections of the stacks; in segments *C*, the PSA-reactive cisternae again are in the ultimate (**b**) or penultimate (**a**) cis position, although the trans saccules are located at the opposite side as compared with their position in segments *A*. **a** ×42500, **b** ×42500

It was the penultimate cis cisterna that showed the most constant and most intense label; strong LCA reaction was also found in limited segments of the transmost cisterna (Fig. 23a). In the immature cells of the crypts, LCA label frequently was distributed across the entire stacks (Fig. 23b).

Some of the goblet cells showed cis-to-trans increase of the LCA label and LCA staining of secretory materials; in other goblet cells, the secretion granules were unstained and the Golgi label restricted to cis/medial cisternae.

Two findings were particularly interesting as regards Golgi architecture:

1. The cisternae reactive for a given lectin frequently corresponded to certain, e.g., cis, medial, trans, or transmost, subsections of the Golgi stacks; reactions, however, were variable within individual cisternae, intensely stained regions alternating with weakly reactive and unreactive regions (e.g., Figs. 18, 19, 20b,

21 a, 23 a, b). This "mosaiclike" staining accounted for the appearance of different patterns within individual stacks. For example, in segments A of the stacks shown in Fig. 23 b, LCA reaction is limited to cisternae of one side; in the neighboring segments of the same stacks (segments B), LCA staining is apparent in all the stacked cisternae.

2. Cisternae with certain lectin-binding characteristics were found in different subsections of the Golgi stacks: in segments A of the stacks shown in Fig. 24, PSA-labeled cisternae are confined to one side; this is presumably the cis side. In segments B, the same PSA-reactive cisternae are located in the medial subsection of the stacks; on both sides, the reactive cisternae are accompanied by unstained (medial and trans?) saccules. In segments C, the PSA-stained cisternae are again limited to one side of the stacks, which presumably is again the cis side; the trans cisternae are located opposite the position in segment A. The PSA-reactive cisternae build up a kind of backbone throughout the stacks.

Both mosaiclike staining and "backbone cisternae" indicate that cisternae with certain lectin-binding characteristics do not necessarily correspond to morphologically defined subsections of the Golgi stacks.

5 Uptake in the Golgi Apparatus of Internalized Molecules

The Golgi apparatus is a main stage in the intracellular routes of a series of molecules taken up by endocytosis and in the recycling pathways of plasma membrane constituents, glycoproteins as well as lipids (for review and further literature, see Brown et al. 1983; Farquhar 1982, 1983, 1985; Geisow 1982; Herzog 1983, 1984; Morré et al. 1984a; Pastan and Willingham 1985; Schwarz and Thilo 1983; Sleigh and Pagano 1984; Steinman et al. 1983; Stenseth and Thyberg 1986; Widnell et al. 1982).

In most studies, the internalized molecules have been recovered in elements of the trans Golgi side; e.g., several ligands subsequent to their uptake by specific or unspecific adsorptive mechanisms can be traced in structures that reside in the transmost position of the stacks, i.e., in rigid lamellae – GERL-like structures, tubular-reticular elements, multivesiculated bodies, and polymorphous vesicles possessing tubular appendices (Fig. 9a–c). Appearance in these elements of internalized ligands may follow dissociation from their receptors in the endosome/receptosome/CURL system (cf. Hopkins 1985; Willingham and Pastan 1980; Geuze et al. 1983a). On the other hand, several of the transmost Golgi elements which contain internalized ligands strikingly resemble those structures defined as constituents of the endosome/receptosome/CURL system, i.e., vacuoles containing vesicular inclusions, polymorphous bodies with tubular extensions (Fig. 9a–c; cf. also Geuze et al. 1985). This finding and others (e.g., acidic milieu, Anderson and Pathak 1985) raise the possibility that some of the manifold elements of the transmost Golgi section correspond to constituents of the endosome/receptosome/CURL system (cf. also pp. 21–22).

It has been noted that the intracellular routes following internalization are different for various conjugates of particular ligands (e.g., peroxidase, ferritin, and colloidal gold conjugates), either involving elements of the Golgi apparatus or not (e.g., Gonatas et al. 1984; van Deurs et al. 1986). In the case of ricin conjugates, it has been shown that partitioning processes that direct some molecules to the Golgi apparatus, but not others, are dependent on the valency of the conjugates (van Deurs et al. 1986).

The distribution in the trans/transmost Golgi structures is different for various internalized molecules; e.g., internalized transferrin has been localized in the tubular elements of the trans Golgi tubular network, whereas epidermal growth factor is concentrated in the coated pits of this system (Hanover et al. 1984; Willingham et al. 1984). It is thought that this distinct location of endocytic substances reflects a targeting to diverse cellular destinations, e.g., to the plasma membrane in the case of transferrin and to lysosomes in the case of epidermal growth factor.

While several endocytic molecules, such as transferrin (Hanover et al. 1984;

Willingham et al. 1984), epidermal growth factor (Hanover et al. 1984; Willingham et al. 1983a), asialoglycoproteins (Deschuyteneer et al. 1984; Geuze et al. 1983b; Haimes et al. 1981), α_2-macroglobulin (Willingham et al. 1980), lysosomal enzymes (Willingham et al. 1981b), peroxidase (Broadwell and Oliver 1983; Pelletier 1973), cholera toxin (Joseph et al. 1979), and various lectins (e.g., Gonatas et al. 1977, 1984; Stieber et al. 1984; van Deurs et al. 1986; Wang et al. 1983), have been recovered almost exclusively in elements of the trans/transmost Golgi subsections (Figs. 9a–c, 10c), others apparently gain access to cisternae of all subsections of the Golgi stacks. For example, internalized dextrans and cationized ferritin (Fig. 10a, b) have been detected in cisternae of each Golgi subsection (Farquhar 1978; Herzog 1984; Herzog and Farquhar 1977; Herzog and Reggio 1980; Ottosen et al. 1980; Thyberg 1980). This not only concerns exogenous molecules taken up by endocytosis, but also recycling plasma membrane molecules. With myeloma cells, it has recently been shown that recycling plasma membrane transferrin receptor not only is taken up into trans cisternae, but gains access also to cis and medial cisternae of the Golgi stacks (Woods et al. 1986). This finding is consistent with the results of another study making use of the Golgi mannosidase I inhibitor deoxymannojirimycin (Snider and Rogers 1986): cellular glycoproteins including plasma membrane transferrin receptor recycle to the sites where Golgi mannosidase I is acting, this presumably being a cis or medial Golgi compartment (Dunphy and Rothman 1983; Kornfeld and Kornfeld 1985).

It is not clear whether internalized molecules may enter the Golgi apparatus at each level of the stacks, i.e., at the cis, medial, trans, and transmost subsections, or enter at one subsection, subsequently being transferred from there to the other subsections. In myeloma cells (Woods et al. 1986), at short time intervals (i.e., 5 min) after internalization, cis and medial cisternae have been found to be the predominant acceptor compartments for recycling plasma membrane transferrin receptor. At later time intervals (i.e., 30 min), the immunoreaction localizing the internalized receptor is shifted to the trans cisternae, cis and medial cisternae still being reactive. In the case of recycling transferrin receptor in erythroleukemia cells (Snider and Rogers 1986), the results indicate that transfer to the sialyltransferase compartment, which presumably corresponds to the transmost subsection of the Golgi stacks (Roth et al. 1985), occurs faster as compared with transport to the Golgi mannosidase I compartment. This may be a result of preferential uptake of transferrin receptor in vesicles that are destined for the sialyltransferase compartment or reflect a greater rate of vesicular transport into the sialyltransferase compartment (Snider and Rogers 1986).

Uptake in Golgi elements is considered as being significant for repair, e.g., resialylation (Regoeczi et al. 1982; Snider and Rogers 1985) of internalized molecules, and for their targeting to diverse cellular locations (for review and further literature, see, e.g., Breitfield et al. 1985; Farquhar 1983, 1985; Morré et al. 1984a; Pastan and Willingham 1985). The findings that internalized plasma membrane molecules may visit not only trans but also cis and medial cisternae of the Golgi stacks (Woods et al. 1986) and recycle to compartments where enzymes of the early steps of glycoprotein processing, such as Golgi mannosidase I (Snider and Rogers 1986), are located, raise the possibility that not only resialylation but more extensive modifications occur during recycling (cf. Woods et al. 1986).

6 Discussion

The Golgi apparatus is a continuous membrane system, stacks of flat cisternae, interconnected by cisternal or tubular-reticular "bridges," representing morphologic subunits.

6.1 Morphologic Subsections – Functional Subcompartments

Morphologically and by means of diverse cytochemical procedures, cis, medial, trans, and transmost subsections may be distinguished in the individual stacks. Furthermore, both biochemical (e.g., Bergeron et al. 1985; Deutscher et al. 1983; Dunphy and Rothman 1983; Elhammer and Kornfeld 1984; Gabel and Bergmann 1985; Goldberg and Kornfeld 1983; Morré et al. 1984b; Pohlmann et al. 1982; Quinn et al. 1983) and morphologic (e.g., Bennett and O'Shaughnessy 1981; Berger and Hesford 1985; Dunphy et al. 1985; Griffiths et al. 1982, 1983; Pavelka and Ellinger 1985; Roth 1984; Roth and Berger 1982; Roth et al. 1985; Sato and Spicer 1982a, b; Tartakoff and Vassalli 1983) findings have shown that functionally different subcompartments exist in the Golgi apparatus (for recent reviews, see Berger 1985; Dunphy and Rothman 1985; Tartakoff 1983a).

How functional subcompartments are arranged in the complex Golgi system, and how functional subcompartments are related to and fit into the morphologic picture of the Golgi subsections, are crucial questions connected with Golgi organization.

6.2 Morphology and Classic Cytochemistry

The polar construction, morphologically apparent, and the differentiated cytochemical reactions as observed in several cell types (osmium, cis; NADPase, medial; TPPase, trans, and AcPase, transmost) probably reflect presence and arrangement of functionally different subcompartments of the Golgi stacks. However, both morphologic appearance and the classic cytochemical reactions are so variable that a common pattern in the organization of Golgi subcompartments cannot be deduced:

1. Several cell types do not show a polar construction of the Golgi stacks; morphologic polarity and cytochemical patterns do not necessarily correspond to each other (e.g., in the duodenal and jejunal absorptive cells,

cf. p. 36). Scanning electron microscopic findings indicate the existence of stacks that are composed of one single, helically wound cisterna (Tanaka et al. 1986).

2. Structures that have been attributed to the trans Golgi section, such as rigid lamellae, may be located in medial sections of the stacks (cf. p. 19).
3. Golgi products do not become packaged exclusively at the trans Golgi side; secretory granules may pinch off from all the cisternae throughout the stacks (cf. also formation of granules in the developing granulocytes, Bainton and Farquhar 1966).
4. Enzyme modulations accompany cell differentiation and occur in response to specific cell stimulation (reviewed in Oliver and Hand 1983). So far, none of the cytochemical enzyme reactions can be considered a reliable marker for labeling certain subsections of the Golgi stacks.

The variability of the classic cytochemical Golgi reactions, like the variations of the lectin-cytochemical patterns, suggests that the arrangement of functionally different subcompartments of the stacks may change; this points to flexibility of the organization of Golgi elements.

6.3 Immunocytochemistry and Lectin Cytochemistry

By means of immunocytochemistry and lectin cytochemistry, subsections have been localized in the Golgi stacks that may be related to definite Golgi functions, e.g., certain steps in the biosynthesis of glycans.

Various enzymes of the glycan-processing system (Hubbard and Ivatt 1981; Kornfeld and Kornfeld 1985; Montreuil 1984), active in the step-by-step addition of terminal sugars to the growing saccharide chains, have been demonstrated immunocytochemically in cisternae of different subsections of the stacks: N-acetyl-glucosaminyltransferase I in medial cisternae (Dunphy et al. 1985) and galactosyl- and sialyltransferase in elements of the trans and transmost subsections (Berger and Hesford 1985; Roth and Berger 1982; Roth et al. 1985; Strous et al. 1983a). Sialyltransferase immunoreactivity has been found particularly concentrated in the transmost tubular-reticular elements of the Golgi stacks of hepatocytes (Roth et al. 1985).

This localization of glycan-processing enzymes, confined to certain subsections of the Golgi apparatus, is in line with some of the lectin-cytochemical patterns: in several cell types, reactions for ConA, which particularly binds with high-mannose N-linked oligosaccharides, are limited to cis/medial cisternae (Pavelka and Ellinger 1985; Tartakoff and Vassalli 1983; Figs. 16, 18a). Binding sites for RCA I (Griffiths et al. 1982; Pavelka and Ellinger 1985; Roth 1983) and WGA (Tartakoff and Vassalli 1983), recognizing terminal galactose and N-acetyl-glucosamine/sialic acid, respectively, predominate in the medial/trans subsections of the stacks (Figs. 17, 18b). LFA, which is specific for sialic acid, particularly labels cisternae of the trans/transmost Golgi subsections (Roth et al. 1984). These differentiated Con A-cis and RCA I/WGA/LFA-trans/transmost reactions, as obtained in some cell types, may mirror the conversion of high-mannose-type N-linked oligosaccharides into complex-type oligosaccharide species.

Furthermore, differentiated Golgi stack labeling is obtained with PNA, HPA, PSA, and LCA: in various types of cells of the gastrointestinal tract, PNA, which recognizes N-acetyl-galactosamine-galactose sequences, has been shown to bind with constituents of the medial Golgi cisternae (Sato and Spicer 1982a, b). HPA, a N-acetyl-galactosamine probe, particularly labels cis cisternae (Pavelka and Ellinger 1985; Roth 1984; Fig. 21). The HPA-cis reaction presumably indicates initial steps in the synthesis of O-linked oligosaccharides (Elhammer and Kornfeld 1984; Roth 1984); HPA reaction of the trans saccules has been considered to reflect the insertion into glycoconjugates of blood group A-specific terminal N-acetyl-galactosamine residues (Roth 1984). PSA and LCA specifically bind to mannose and glucose; for high-affinity binding to glycopeptides, a fucose residue attached to the asparagine-linked N-acetyl-glucosamine is essential (Debray et al. 1981; Kornfeld et al. 1981). Hence, the reactions obtained with PSA and LCA may be considered as chiefly indicating presence of glycopeptides containing core-fucosylated N-linked oligosaccharides. The labeling patterns show that PSA- and LCA-binding molecules are particularly concentrated in cisternae of the cis and medial subsections of the Golgi stacks, the penultimate cis cisterna being the favored one.

The differentiated immunostaining of glycan-processing enzymes, as well as the differentiated lectin reactions obtained in the Golgi apparatus of some cell types, suggest a compartmentalized organization of terminal steps in the biosynthesis of glycans, with the sequence of steps being oriented from the cis to the trans side of the stacks (cf. Tartakoff 1983a; Dunphy and Rothman 1985; Rothman 1985; cf. also Berger 1985; Roth et al. 1985).

Newly synthesized molecules are exported out of the endoplasmic reticulum by transport vesicles and may enter the Golgi apparatus at the cis side of the stacks. N-linked oligosaccharides, after the cleaving-off of mannose residues, may acquire N-acetyl-glucosamine in medial cisternae, thus becoming acceptors for galactose that may be inserted in the trans cisternae; addition of the terminal sialic acid may occur in the transmost Golgi section.

This model implies a strict cis-to-trans orientation of the sequence of processes and, hence, requires strict cis-to-trans transport across the stacks of molecules to be processed. The majority of molecules traversing the Golgi complex and demonstrated cytochemically has been detected in all the stacked cisternae; this suggests that the entire stack of cisternae is involved in the pathways of newly synthesized secretory, lysosomal, and membrane proteins. These results, however, cannot answer the questions as to whether each molecule touches all cisternae or possibly only some cisternae or only one cisterna of the stacks, nor as to where the molecules enter and exit from the Golgi apparatus. Transport oriented from one side, which presumably is the cis side, to the other side, presumably representing the trans side of the stacks, has been indicated by the results of time sequence studies concerning traffic of virus glycoproteins (Bergmann and Singer 1983; Saraste and Kuismanen 1984). In the majority of the stacks (70%, Bergmann and Singer 1983), the immunoreactions localizing the respective virus proteins first occur in the cisternae of one side, which presumably correspond with cis cisternae, and after a time interval of 2–3 min are uniformly distributed across all the stacked cisternae. From the system studied (G proteins of the vesicular stomatis virus in Chinese hamster ovary cells), it has been calculated that the average transit time to traverse the entire stack of cisternae is in the order of 2 min, and the average transit time between one saccule and the next one is in the order of 0.3 min.

In another investigation, cisternae showing immunoreactions for virus proteins that traverse the stacks have been found interposed between unreactive saccules (Saraste and Kuismanen 1984). Patterns that show reactive Golgi cisternae alternating with unreactive saccules have been observed also by OsKI labeling (Locke and Huie 1983); similar distribution of reactions is apparent after ConA (Fig. 20a) and RCA I (Fig. 20b) staining. With respect to OsKI labeling, it has been suggested that the "unstained" Golgi cisternae may be built by fusion of empty transition vesicles returning from the Golgi complex to the endoplasmic reticulum (cf. Locke and Huie 1983).

Several recent studies indicate that endoplasmic reticulum proteins are retained in this compartment and do not touch Golgi elements (Brands et al. 1985; Yamamoto et al. 1985). Immunoreactions for the oligosaccharide-trimming enzyme glucosidase II have been localized in cisternae of the rough and smooth endoplasmic reticulum and in transitional elements but found excluded from Golgi cisternae (Lucocq et al. 1986).

Different membrane glycoproteins are exported from the endoplasmic reticulum and transported to the Golgi apparatus at different rates (Fitting and Kabat 1982; Gorvel et al. 1986; Hauri et al. 1985); furthermore, the migration rates from the endoplasmic reticulum to the Golgi complex and across the Golgi stacks are greatly different for diverse, glycosylated as well as nonglycosylated, secretory proteins (Fries et al. 1984; Lodish et al. 1983; Scheele and Tartakoff 1985). The sum of these findings indicates that the endoplasmic reticulum-to-Golgi transport is a selective process that may be receptor-mediated, similar to receptor-mediated endocytosis and receptor-mediated transport established for the Golgi-to-lysosome route of lysosomal enzymes. It has been suggested that newly synthesized molecules may contain a series of "signals" (oligosaccharides, amino acid sequences, cytoplasmically oriented sequences in the case of membrane proteins) by which the routes to the cellular destinations as well as transport kinetics are determined (e.g. Doyle et al. 1985; Fitting and Kabat 1982; Fries et al. 1984; Guan et al. 1985; Hauri et al. 1985; Lodish et al. 1983; Rose and Bergmann 1983; Scheele and Tartakoff 1985). Associations of secretory molecules and Golgi membranes have been shown in pancreatic B cells (Orci et al. 1984a); here, immunocytochemical label for proinsulin has been found associated with membranes of Golgi cisternae. By contrast, in the secretory granules, the immunoreactions predominate over the dense cores. The proinsulin-Golgi membrane associations are considered as possibly reflecting the presence of specific proinsulin binding sites, which may play a role in the routing of the proinsulin molecules and/or be related to the proteolytic processing of the peptides (Orci et al. 1984a).

Differences of the pH between the diverse subsections of the Golgi stacks (cf. pp. 17–19) are considered as playing a role in the regulation of receptor-mediated transport of molecules across the Golgi stacks. Molecules, upon entering the acidic milieu in the trans Golgi cisternae, may dissociate from the receptor, thus permitting the receptors to recycle back to medial/cis cisternae (Anderson and Pathak 1985).

It has not yet been clarified whether transport across the stacks of cisternae of membrane constituents as well as cisternal contents occurs according to a general principle. Movement of the cisternae "as a whole" (Morré and Ovtracht 1977; Morré et al. 1979; see also Brown and Willison 1977; Locke and Huie 1983; Matsuura and Tashiro 1979; McFadden and Melkonian 1986; Saraste and Kuismanen 1984), transport via intercisternal connections (Bracker et al. 1971; Tanaka et al. 1986), and transport via vesicles pinching off from and fusing with the lateral extremities of the cisternae (e.g., Rothman et al. 1984a, b) have been suggested. In the latter model, the cis-to-trans traffic of newly synthesized molecules is assumed to be accompanied by membrane recycling via vesicles traveling from the trans to medial and cis cisternae and from the cis Golgi side to the endoplasmic reticulum. The numerous small vesicles frequently apparent near the poles and between the stacks of cisternae may represent such an intercisternal transport system. Vesicular transport between different functional compartments of different Golgi stacks has recently been demonstrated by fascinating investigations with hybrids (Rothman et al. 1984a, b) and by using a cellfree in vitro system (Balch et al. 1984a, b; Braell et al. 1984a; Pâquet et al. 1986; Wattenberg et al. 1986). Membrane constituents (virus glycoproteins) traversing the Golgi complex leave the original stack and change over to others; in the systems studied, the stacks functioning as donors and those containing the acceptor compartment derive from different cells.

Energy is required for multiple steps along the pathways of secretory and membrane molecules; these include exit out of the endoplasmic reticulum (Jamieson and Palade 1968; Novick et al. 1981; Tartakoff and Vassalli 1977), transport from cis to medial cisternae of the Golgi stacks (Balch et al. 1984b), as well as exit out of the stack of Golgi cisternae (Novick et al. 1981; Tartakoff 1986). In pancreatic acinar cells, in the absence of ATP production, no budding transitional elements of the endoplasmic reticulum have been found (Tartakoff 1986); furthermore, inhibition of ATP production is accompanied by proliferation of rigid lamellae and an increase in the number of coated vesicles (Tartakoff 1986). The latter finding possibly is related to ATP requirement for removal of clathrin from coated membranes (Braell et al. 1984b; Schlossman et al. 1984; Schmid et al. 1984).

For selected steps along cellular transport routes, temperature requirements have been identified; this concerns both cells with constitutive as well as regulated pathways, although critical temperatures and the sites at which vesicular transport is interrupted differ between diverse cell types studied. For example, endoplasmic reticulum-to-Golgi-to-plasma membrane transport of virus glycoproteins has been studied in tissue culture cells belonging to cells of the constitutive type (Bergmann and Singer 1983; Fuller et al. 1985; Griffiths et al. 1985; Matlin and Simons 1983; Saraste and Kuismanen 1984): for transport out of the endoplasmic reticulum and for traffic across the Golgi stacks 15° and 20° C, respectively, have been found to be permissive temperatures. At 20° C, virus proteins are arrested in extensive tubular-reticular elements in the trans-most position of the Golgi stacks (Griffiths et al. 1985); transport to the cell surface proceeds upon warming of the cells to 32° C.

In pancreatic acinar cells, representing cells of regulated type of secretion, transport out of the endoplasmic reticulum of secretory proteins occurs at >10° C, whereas traffic across the stacked cisternae and exit out of the stacks requires >22° C (Tartakoff 1986). At 16° C (Saraste et al. 1986), secretory proteins can reach a first station that consists of cis Golgi elements; at 20° C, a second compartment is entered by the secretory proteins, probably corresponding to medial Golgi elements. It appears that different temperature thresholds exist for different steps of the secretory pathway, possibly being connected with differences in membrane fluidity. Interruption of endoplasmic reticulum-to-Golgi traffic at $<10°$ C is accompanied by an increase in number and occurrence of long and complex forms of transitional elements (Tartakoff 1986). At 16° C, the number of small vesicles at the cis and trans Golgi side is increased; the Golgi cisternae are swollen and partially fragmented, and condensing vacuoles appear reduced in number and size (Saraste et al. 1986).

Furthermore, cAMP may play a role in connection with Golgi functions. By immunocytochemistry, regulatory and catalytic subunits of cAMP-dependent protein kinase type II have been shown concentrated in the Golgi area (DeCamilli et al. 1986; Nigg et al. 1985a); this possibly reflects importance in mediating cAMP effects on Golgi-associated processes.

It is not yet established which mechanisms exist in the Golgi complex for discriminating between endogenous Golgi proteins and proteins traversing the organelle, and for maintaining endogenous Golgi proteins in the respective subsections of the stacks. In the case of galactosyltransferase in HeLa cells, it has been shown that the soluble protein that is secreted has a lower molecular weight as compared with the Golgi-associated protein (Strous and Berger 1982) and that the latter polypeptide is located within the membranes of the Golgi cisternae (Strous et al. 1983a). Interactions of the anchoring peptide with the specific membrane environment have been considered probably to be related to the residency of galactosyltransferase in the Golgi complex (Strous et al. 1983a). Recently, virus glycoproteins (Uukuniemi virus) have been found to be retained in the Golgi complex, thus behaving like Golgi-specific proteins (Gahmberg et al. 1986).

The model of "compartmentalized cis-to-trans-oriented Golgi organization," implying definite sites for entrance into (cis) and exit out (trans) of the Golgi stacks of newly synthesized molecules, as well as strict orientation of the sequence of events from cis to trans, is fascinating; in such an organization, each stack of cisternae could be seen as a functional subunit of the Golgi complex, being itself composed of functional subcompartments. In this model, functional subunits would correspond to morphologic subunits (= the Golgi stacks), and functional subcompartments essentially correspond to the morphologically defined cis – medial – trans – transmost subsections of the stacks. However, several findings are difficult to reconcile with a strict cis-to-trans-oriented organization of Golgi subcompartments.

6.3.1 Exit Sites

The trans/transmost Golgi sections appear to be the main exit sites of Golgi products from the Golgi apparatus; in several cell types, however, secretory granules can be observed to form from the rims of cis and medial cisternae as well. In the jejunal absorptive cells, lipid particle-containing secretory vesicles pinch off from the saccules (presumably cis) that are located opposite those that exhibit morphologic and cytochemical trans features (Pavelka and Ellinger 1981a). In the developing granulocytes, the two types of granules (azurophil and specific) derive from opposite faces of the Golgi stacks (Bainton and Farquhar 1966). The tubules which constitute "interstack bridges" are the sites where secretion granules predominantly are formed in atrial cardiocytes (Rambourg et al. 1984). The cis-localized immunoreactions of the mannose-6-phosphate receptor suggest that lysosomal enzymes leave the Golgi apparatus at the cis side of the stacks (Brown and Farquhar 1984a).

Hence, it appears that different molecules are exported from the Golgi apparatus at different subsections of the stacks, including trans as well as cis cisternae. This poses questions as to the sites in the Golgi stacks at which processing takes place of molecules that exit from cis cisternae; e.g., it is known that lysosomal enzymes contain not only high mannose but also complex-type and hybrid-type oligosaccharides (Hasilik and von Figura 1981). If exported from the Golgi complex at the cis side, where do they acquire the terminal sugars? If terminal glycosyltransferases are limited to the trans side of the stacks (Berger and Hesford 1985; Roth and Berger 1982; Roth et al. 1985; Strous et al. 1983a), lysosomal enzymes, after having acquired terminal sugars in the trans cisternae, would subsequently have to be transported to the cis side of the stacks.

6.3.2 Variability of the Golgi Architecture

Variability of Golgi morphology and enzyme-cytochemical patterns suggests that the organization of the Golgi apparatus is flexible. The lectin-cytochemical reactions also show variations. The clear-cut ConA-cis and RCA I/LFA-trans reactions, which may mirror a cis-to-trans-oriented conversion of high mannose to complex-type species of N-linked oligosaccharides, is missing in several cell types (Griffiths et al. 1982; Hedman et al. 1986; Pavelka and Ellinger 1986b–d; Roth et al. 1986); e.g., in the embryonic pancreatic acinar cells, both ConA and RCA I binding can be found in the cisternae of all Golgi subsections (Figs. 19a–c, 20a, b). LFA binding reactions are extended across the majority of the stacked Golgi saccules in the absorptive cells of the colon (Roth et al. 1986); in 3T3 fibroblasts, LFA reactions have been demonstrated in most of the stacked cisternae, only one cisterna at one side being unreactive (Hedman et al. 1986).

The RCA I reactions of trans Golgi cisternae as appear in some cell types correspond to the trans localization of galactosyltransferase as demonstrated by immunocytochemistry (Roth and Berger 1982; Strous et al. 1983a); this might mean that RCA I labeling indicates the site of insertion of galactose into the growing saccharide chains. RCA I reaction has been considered to be a marker for characterizing elements of the trans side of the Golgi stacks (Bergmann and Singer 1983; Ratcliffe et al. 1985). As regards an interpretation of the RCA I reactions of medial and cis cisternae as observed in baby hamster kidney cells (Griffiths et al. 1982), intestinal absorptive cells (Pavelka and Ellinger 1986c), and embryonic acinar cells (Pavelka and Ellinger 1986b, c), three main points have to be considered:

1. RCA I reactions of cis and medial cisternae may indicate that galactose-transferring enzymes are not uniquely found in cisternae of the trans side of the Golgi stacks but are also present in medial and cis subsections. Different galactosyltransferase species catalyzing the transfer to diverse acceptor mole-

cules (Kaplan and Hechtman 1984) may be located in different subsections of the Golgi stacks. The RCA I patterns are in agreement with recent biochemical as well as immunocytochemical findings. In hepatocytes, transfer of galactose to endogenous protein acceptors has been found to be pronounced in medial and cis Golgi elements (Bergeron et al. 1982a; Morré et al. 1983). Three terminal glycosyltransferases, including galactosyltransferase, have been demonstrated in three different liver cell Golgi subfractions (Bretz et al. 1980), differences of the specific activities existing between the individual subfractions. Free flow electrophoresis studies (Morré et al. 1983, 1984b) have indicated that galactosyltransferase and sialyltransferase occupy the same cisternae and exhibit a gradient of increasing activity toward the trans side. By means of a cytochemical method that couples uridine diphosphate formation and nicotinamide adenine dinucleotide reduction, galactosyltransferase has been demonstrated in cisternae of Golgi apparatus isolated from rat liver showing the most intense reaction in the medial region of the stacks (Matyas and Morré 1983).

Immunocytochemically, in contrast to other cell types, in absorptive cells of the large intestine sialyltransferase and blood group A α 1,3 N-acetylgalactosaminyltransferase have been found distributed throughout all the stacked cisternae, except the cismost one (Roth et al. 1986).

These immunocytochemical findings, together with the lectin-cytochemical patterns obtained with RCA I (Figs. 19, 20b) and LFA (Hedman et al. 1986; Roth et al. 1986) and enzyme-cytochemical results (Fig. 15), show that characteristics which in a series of cells have been attributed to trans cisternae can be apparent in all the stacked saccules; this may reflect either lack of compartmentation (Fig. 25b/II) or be due to proportionally different distribution of subcompartments in the different Golgi segments (Fig. 25b/III-IV and V–VI). The appearance of different reaction patterns in stacks located side by side, such as shown with RCA I (Fig. 19b), may reflect that Golgi subcompartments can be distributed in the stacks at heterogeneous proportions.

2. RCA I staining of cis and medial cisternae may be the result of the transport to cis and medial cisternae of glycoconjugates that have acquired galactose in the trans cisternae; this implies trans-to-cis-directed traffic that runs in the opposite direction as compared with the cis-to-trans transport usually proposed for newly synthesized molecules. Such trans-to-cis-directed traffic may be significant for molecules that are packaged at the cis side of the Golgi stacks (e.g., Brown and Farquhar 1984a).

3. The RCA I label may not be the exclusive result of binding with newly synthesized molecules; internalized molecules and membrane constituents retrieved from the cell surface or other cellular compartments and taken up in Golgi elements may also contribute to the RCA I reactions. However, in the cell types studied so far, internalized RCA I has been taken up almost exclusively in cisternae of the trans/transmost Golgi subsections (e.g., van Deurs et al. 1986).

6.3.3 Mosaiclike Patterns

The reactions obtained within the individual cisternae by means of lectin labeling are variable, such as has also been shown by other techniques (periodic acid

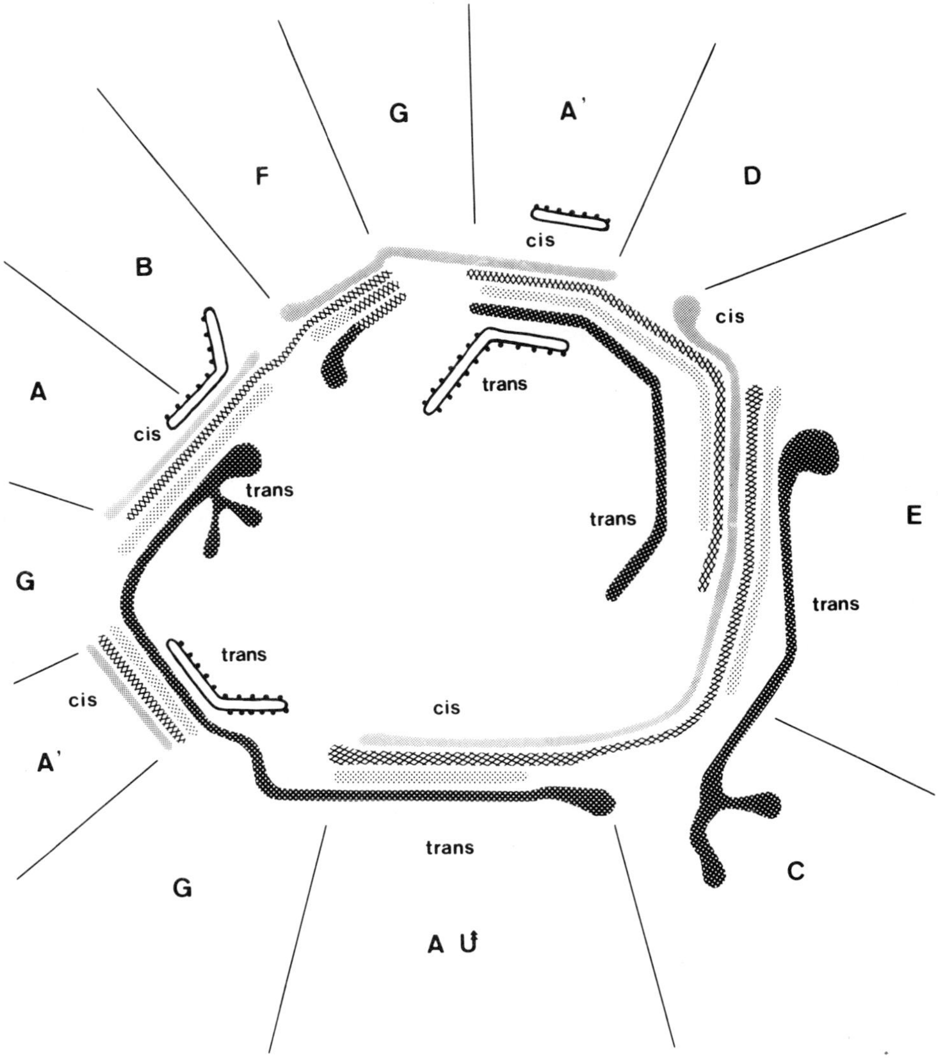

Fig. 25 a, b. A summary of possible arrangements of the main subcompartments of the Golgi apparatus as revealed from morphologic/cytochemical images. **a** The arrangement of subcompartments may differ between different Golgi segments (*A–F*).
A, cis, medial, trans, transmost; *A'*, cis, medial, trans, transmost + endoplasmic reticulum; *A* **U**, cis, medial, trans, transmost, in reverse position to the stack in segment *A*; cf. Fig. 6a; *B*, cis, medial, trans; *C*, cis, medial, transmost; *D*, medial, trans, transmost; *E*, transmost, trans, medial, cis, medial, trans, transmost, the cis cisterna = backbone cisterna; cf. Fig. 24a, b; *F*, mosaic staining; cf. Figs. 21a, 23b; *G*, interstack bridge; cf. Figs. 6a, b, 20b.
The subcompartments do not necessarily correspond to the morphologic subsections of the stacks; e.g., in segment *B*, the trans compartment is in the transmost position; in segment *D*, the medial subcompartment is in the cismost position; in segment *E*, the cis subcompartment corresponds to the medial subsection. **b** The proportional distribution of subcompartments may differ between different segments (*I–VI*) of the Golgi apparatus.
I, compartmentalized organization, functional subcompartments are equally distributed and correspond to the morphologic subsections; *II*, lack of compartmentation, characteristics of all subcompartments are present in all the stacked cisternae; *III/IV*, predominance of cis and medial subcompartments; *V/VI*, predominance of trans and transmost subcompartments

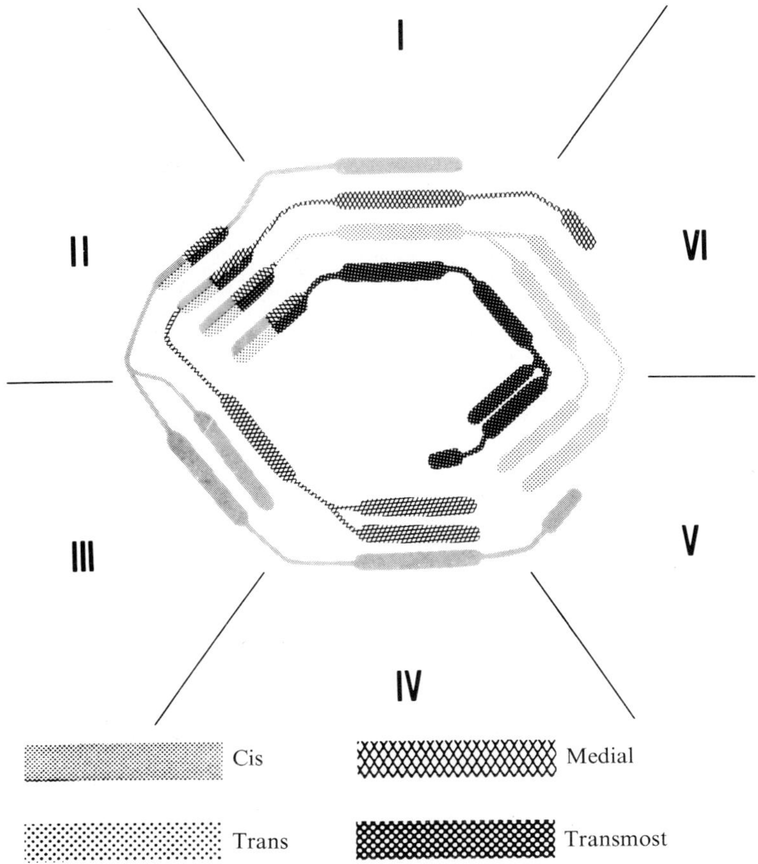

Fig. 25b

silver staining, Ovtracht and Thiery 1972; phosphotungstic acid staining at low pH, Morré and Mollenhauer 1983). With each of the lectins tested, binding sites were found concentrated in certain limited segments of the Golgi cisternae, neighboring segments being faintly labeled or unreactive (e.g., Figs. 18, 19, 20b, 21a, 23a, b); this indicates that functionally different domains may exist within the individual cisternae.

The mosaiclike staining of cisternae accounts for the appearance of different lectin reaction patterns within the individual Golgi stacks: as demonstrated in Fig. 23b, some segments of the stacks may show the reaction confined to one side, while other segments may exhibit the same reactions distributed across all the stacked cisternae. In line with the results of affinity separation techniques (Ito and Palade 1978; reviewed in Farquhar and Palade 1981), the mosaiclike patterns suggest that functional subdissection of the Golgi stacks exists not only along the cis-to-trans axis but also along the pole-to-pole axis of the individual cisternae.

6.3.4 "Backbone Cisternae"

Cisternae with certain lectin binding characteristics may constitute different subsections of the Golgi stacks. As shown for PSA in Fig. 24, cisternae with the same binding characteristics represent the cis subsection in one segment of the stacks, the medial subsection in the neighboring segment, and again the cis subsection in the subsequent segment. The reactive cisternae form a kind of backbone throughout the stack, which is accompanied by unreactive (medial and trans?) cisternae at both sides.

"Cis-to-trans-extended" reactions (Figs. 19, 20, 21a; Griffiths et al. 1982; Hedman et al. 1986; Pavelka and Ellinger 1986b–d; Roth et al. 1986), mosaic-like staining (Figs. 18, 19, 20b, 21a, 23a, b), and backbone cisternae (Fig. 24) indicate that Golgi segments with certain lectin-binding characteristics, which may correspond to certain functional subcompartments, can be distributed within the Golgi apparatus irrespective of the morphologic subdivision into cis-medial-trans-transmost subsections.

The diagrams shown in Fig. 25a, b, give a synopsis of possible arrangements of the main subcompartments of the Golgi apparatus as revealed from morphologic/cytochemical studies.

7 Summary

The Golgi apparatus is a crossroad in intracellular traffic; it is a crucial stage in the pathways of newly synthesized as well as internalized and recycling molecules. Having been transported to the Golgi apparatus by vesicular carriers or via membrane continuities, molecules are taken up into and possibly traverse the Golgi complex. Elements of the Golgi apparatus are the sites where molecules of various origin and destination are subjected to multiple modifications, such as proteolytic cleavage, glycosylation, phosphorylation, and sulfation, and from where they subsequently are targeted to diverse cellular locations.

Being a continuous membrane system, the Golgi apparatus is composed of subunits: flat cisternae are arranged in parallel forming the characteristic "Golgi stacks" that are interconnected by cisternal or tubular-reticular "bridges." At the stacks, morphologically four subsections, the cis, medial, trans, and transmost subsections, can be discriminated, and both morphologic and biochemical findings indicate a compartmentalized Golgi organization. How functional subcompartments are arranged in the complex organelle is one of the crucial questions connected with Golgi organization.

In several kinds of cells, functional subcompartments correspond to definite, viz., cis, medial, trans, and transmost, subsections of the stacks; the sequence of processes, e.g., subsequent steps in the synthesis of glycans, appears to be directed from the cis to the trans side. Evidence for a clear-cut cis-to-trans-oriented subcompartmentation of the Golgi stacks has so far been provided for a limited number of cell types, such as goblet cells, but cannot be seen as a general principle of Golgi organization. As revealed by various techniques, functional subcompartments can be distributed in the Golgi apparatus irrespective of the morphologic subdivision in cis-medial-trans-transmost subsections.

References

Alroy J, Merk FB, Morré DJ, Weinstein RS (1982) Membrane differentiation in the Golgi apparatus of mammary urinary bladder epithelium. Anat Rec 203:429–440

Amos WB, Grimstone AV (1968) Intercisternal material in the Golgi body of Trichomonas. J Cell Biol 38:466–471

Anderson RGW, Pathak RK (1985) Vesicles and cisternae of the trans Golgi apparatus of human fibroblasts are acidic compartments. Cell 40:635–643

Angermüller S, Fahimi HD (1984) Cytochemical localization of β-NADPase in rat hepatocytes and Kupffer cells. J Histochem Cytochem 32:541–546

Bainton DF, Farquhar MG (1966) Origin of granules in polymorphonuclear leukocytes. Two types derived from opposite faces of the Golgi complex in developing granulocytes. J Cell Biol 28:277–301

Bainton DF, Farquhar MG (1968) Differences in enzyme content of azurophil and specific granules in polymorphonuclear leukocytes. II. Cytochemistry and electron microscopy of bone marrow cells. J Cell Biol 39:299–317

Balch WE, Dunphy WG, Braell WA, Rothman JE (1984a) Reconstitution of the transport of protein between successive compartments of the Golgi measured by the coupled incorporation of N-acetylglucosamine. Cell 39:405–416

Balch WE, Glick BS, Rothman JE (1984b) Sequential intermediates in the pathway of intercompartmental transport in a cell-free system. Cell 39:525–536

Banerjee D, Redman CM (1984) Biosynthesis of high density lipoprotein by chicken liver: Conjugation of nascent lipids with apoprotein A1. J Cell Biol 99:1917–1926

Banerjee D, Manning CP, Redman CP (1976) The in vivo effect of colchicine on the addition of galactose and sialic acid to rat hepatic serum glycoproteins. J Biol Chem 251:3887–3892

Barka (1964) Electron histochemical localization of acid phosphatase activity in the small intestine of mouse. J Histochem Cytochem 12:229–238

Baron R, Neff L, Louvard D, Courtoy PJ (1985) Cell-mediated extracellular acidification and bone resorption: Evidence for a low pH in resorbing lacunae and localization of a 100-kD lysosomal membrane protein at the osteoclast ruffled border. J Cell Biol 101:2210–2222

Barr R, Safranski K, Sun IL, Crane FL, Morré DJ (1984) An electrogenic proton pump associated with the Golgi apparatus of mouse liver driven by NADH and ATP. J Biol Chem 259:14064–14067

Beaudoin AR, Grondin G, Lord A, Roberge M, St-Jean P (1983) The origin of the zymogen granule membrane of the pancreatic acinar cell as examined by ultrastructural cytochemistry of acid phosphatase, thiamine pyrophosphatase, and ATP-diphosphohydrolase activities. Eur J Cell Biol 29:218–225

Beguinot L, Lyall RM, Willingham MC, Pastan I (1984) Down-regulation of the epidermal growth factor receptor in KB cells is due to receptor internalization and subsequent degradation in lysosomes. Proc Natl Acad Sci USA 81:2384–2388

Bendayan M (1984) Concentration of amylase along its secretory pathway in the pancreatic acinar cell as revealed by high resolution immunocytochemistry. Histochem J 16:85–108

Bendayan M, Roth J, Perrelet A, Orci L (1980) Quantitative immunocytochemical localization of pancreatic secretory proteins in subcellular compartments of the rat acinar cells. J Histochem Cytochem 28:149–160

Bennett G (1984) Role of the Golgi complex in the secretory process. In: Canin M (ed) Cell biology of the secretory process. Karger, Basel, pp 102–147

Bennett G, Leblond CP (1977) Biosynthesis of the glycoproteins present in plasma membrane, lysosomes and secretory materials, as visualized by radioautography. Histochem J 9:393–417

Bennett G, O'Shaughnessy D (1981) The site of incorporation of sialic acid residues into glycoproteins and the subsequent fate of the molecules in various rat and mouse cell types as shown by radioautography after injection of 3H-N-acetylmannosamine. I. Observations in hepatocytes. J Cell Biol 88:1–15

Bennett G, Leblond CP, Haddad A (1974) Migration of glycoprotein from the Golgi apparatus to the surface of various cell types as shown by radioautography after labeled fucose injection into rat. J Cell Biol 60:258–284

Bennett G, Parsons S, Carlet E (1984a) Influence of colchicine and vinblastine on the intracellular migration of secretory and membrane glycoproteins. I. Inhibition of glycoprotein migration in various rat cell types as shown by light microscope radioautography after injection of 3H-fucose. Am J Anat 170:521–530

Bennett G, Carlet E, Wild G, Parsons S (1984b) Influence of colchicine and vinblastine on the intracellular migration of secretory and membrane glycoproteins. III. Inhibition of intracellular migration of membrane glycoproteins in rat intestinal columnar cells and hepatocytes as visualized by light and electron microscope radioautography after 3H-fucose injection. Am J Anat 170:545–566

Berger EG (1984) The functional organization of the Golgi apparatus. INSERM 1984 Vol 126:111–122

Berger EG (1985) Mini-review: How Golgi-associated glycosylation works. Cell Biol Int Rep 9:407–417

Berger EG, Hesford FJ (1985) Localization of galactosyl and sialyltransferase by immunofluorescence: Evidence for different sites. Proc Natl Acad Sci USA 82:4736–4739

Bergeron JJM, Rachubinski RA, Sikstrom RA, Posner BI, Paiement J (1982a) Galactose transfer to endogenous acceptors within Golgi fractions of rat liver. J Cell Biol 92:139–146

Bergeron JJM, Kotwal FJ, Levine G, Bilan B, Rachubinski K, Hamilton M, Shore GC, Ghosh HP (1982b) Intracellular transport of the transmembrane glycoprotein G of vesicular stomatitis virus through the Golgi apparatus as visualized by electron microscopic radioautography. J Cell Biol 94:36–41

Bergeron JJM, Paiement J, Khan MN, Smith CE (1985) Terminal glycosylation in rat hepatic Golgi fractions: Heterogenous locations for sialic acid and galactose acceptors and their transferases. Biochim Biophys Acta 821:393–403

Bergmann JE, Singer SJ (1983) Immunoelectron microscopic studies of the intracellular transport of the membrane glycoprotein (G) of vesicular stomatitis virus in infected Chinese hamster ovary cells. J Cell Biol 97:1777–1787

Bergmann JE, Tokuyasu KT, Singer SJ (1981) Passage of an integral membrane protein, the vesicular stomatitis virus glycoprotein, through the Golgi apparatus en route to the plasma membrane. Proc Natl Acad Sci USA 78:1746–1750

Berlin RD, Caron JM, Oliver JM (1979) Microtubules and the structure and function of cell surfaces. In: Roberts K, Hyams JS (eds) Microtubules. Academic, London, pp 443–485

Bernhard W, Avrameas S (1971) Ultrastructural visualization of cellular carbohydrate components by means of concanavalin A. Exp Cell Res 64:232–236

Blok J, Ginsel LA, Mulder-Stapel AA, Onderwater JJM, Daems WT (1981) The effect of colchicine on the intracellular transport of 3H-fucose-labelled glycoproteins in the absorptive cells of cultured human small intestinal tissue. Cell Tiss Res 215:1–12

Blomfield J, Simson JAV, Martinez AM, Martinez JR (1983) Ultrastructural responses by Golgi apparatus of rat submandibular gland to β-adrenergic, α-adrenergic, and cholinergic stimulation. Exp Mol Path 38:170–182

Bogart BI (1975) Secretory dynamics of the rat submandibular gland. An ultrastructural and cytochemical study of the isoproterenol-induced secretory cycle. J Ultrastr Res 52:139–155

Boutry J-M, Novikoff AB (1975) Cytochemical studies on Golgi apparatus, GERL, and lysosomes in neurons of dorsal root ganglia in mice. Proc Natl Acad Sci USA 72:508–512

Bracker C, Grove SN, Heintz CE, Morré DJ (1971) Continuities between endomembrane components in hyphae of *Pythium* spp. Cytobiol 4:1–8

Braell WA, Balch WE, Dobbertin DC, Rothman JE (1984a) The glycoprotein that is transported between successive compartments in the Golgi in a cell-free system resides in stacks of cisternae. Cell 39:511–524

Braell WA, Schlossman DM, Schmid SL, Rothman JE (1984b) Dissociation of clathrin coats coupled to the hydrolysis of ATP: Role of an uncoating ATPase. J Cell Biol 99:734–741

Brands R, Slot JW, Geuze HJ (1983) Albumin localization in rat liver parenchymal cells. Eur J Cell Biol 32:99–107

Brands R, Snider MD, Hino Y, Park SS, Gelboin HV, Rothman JE (1985) Retention of membrane proteins by the endoplasmic reticulum. J Cell Biol 101:1724–1732

Breitfeld PP, Simmons CF Jr, Strous GJAM, Geuze HJ (1985) Cell biology of the asialoglycoprotein receptor system: A model of receptor-mediated endocytosis. Int Rev Cytol 97:47–95

Bretz R, Bretz H, Palade GE (1980) Distribution of terminal glycosyltransferases in hepatic Golgi fractions. J Cell Biol 84:87–101

Broadwell RD, Cataldo AM (1983) The neuronal endoplasmic reticulum: Its cytochemistry and contribution to the endomembrane system. I. Cell bodies and dendrites. J Histochem Cytochem 31:1077–1088

Broadwell RD, Oliver C (1981) The Golgi apparatus, GERL, and secretory granule formation within neurons in the hypothalamo-neurohypophyseal system of control and hyperosmotically stressed mice. J Cell Biol 90:474–484

Broadwell RD, Oliver C (1983) An enzyme cytochemical study of the endocytic pathways of the mouse in vivo. J Histochem Cytochem 31:325–335

Broadwell RD, Oliver C, Brightman MW (1979) Localization of neurophysin within organelles associated with protein synthesis and packaging in the hypothalamo-neurohypophyseal system: An immunocytochemical study. Proc Natl Acad Sci USA 76:5999–6003

Brodie DA (1981) Bead rings at the endoplasmic reticulum-Golgi complex boundary: Morphological changes accompanying inhibition of intracellular transport of secretory proteins in arthropod fat body tissue. J Cell Biol 90:92–100

Brodie DA (1982a) Golgi complex beads in vertebrates and their relationship with clathrin coats. Tiss Cell 14:253–262

Brodie DA (1982b) The arrangement of the Golgi complex beads is not controlled by the cytoskeleton. Tiss Cell 14:263–271

Brown JC, Hunt RC (1978) Lectins. Int Rev Cytol 52:277–349

Brown MS, Anderson RGW, Goldstein JL (1983) Recycling receptors: The roundtrip itinerary of migrant membrane proteins. Cell 32:663–667

Brown RM Jr, Willison JHM (1977) Golgi apparatus and plasma membrane involvement in secretion and cell surface deposition with special emphasis on cellulose biogenesis. In: Brinkley BR, Porter KR (eds) International Cell Biology 1977, Rockefeller University Press, New York, pp 267–283

Brown WJ, Farquhar MG (1984a) The mannose-6-phosphate receptor for lysosomal enzymes is concentrated in cis Golgi cisternae. Cell 36:295–307

Brown WJ, Farquhar MG (1984b) Accumulation of coated vesicles bearing mannose-6-phosphate receptors for lysosomal enzymes in the Golgi region of I-cell fibroblasts. Proc Natl Acad Sci USA 81:5135–5139

Brown WJ, Constantinescu E, Farquhar MG (1984) Redistribution of mannose-6-phosphate receptors induced by tunicamycin and chloroquine. J Cell Biol 99:320–326

Burchanowski BJ, Hogue-Angeletti R, Stieber A, Gonatas J, Gonatas NK (1982) In "undifferentiated" PC 12 cells, GERL is not segregated from the Golgi apparatus. J Neurocytol 11:323–333

Burke B, Griffiths G, Reggio H, Louvard D, Warren G (1982) A monoclonal antibody against a 135-K Golgi membrane protein. EMBO J 12:1621–1628

Burke B, Walter C, Griffiths G, Warren G (1983) Viral glycoproteins at different stages of intracellular transport can be distinguished using monoclonal antibodies. Eur J Cell Biol 31:315–324

Buschman RJ (1983) Morphometry of the small intestinal enterocytes of the fasted rat and the effects of colchicine. Cell Tiss Res 231:289–299

Busson-Mabillot S, Chambaut-Guérin A-M, Ovtracht L, Muller P, Rossignol B (1982) Microtubules and protein secretion in rat lacrimal glands: Localization of short-term effects of colchicine on the secretory process. J Cell Biol 95:105–118

Capasso JM, Hirschberg CB (1984) Mechanisms of glycosylation and sulfation in the Golgi apparatus: Evidence for nucleotide sugar/nucleoside monophosphate and nucleotide sulfate/nucleoside monophosphate antiports in the Golgi apparatus membrane. Proc Natl Acad Sci USA 81:7051–7055

Carasso N, Ovtracht L, Favard P (1971) Observation en microscopie électronique à haute tension, de l'appareil de Golgi sur coupes de 0.5 à 5 µm d'èpaisseur. CR Acad Sci Paris Ser D 273:876–879

Cataldo AM, Broadwell RD (1982) Morphological and cytochemical associations of the endoplasmic reticulum (ER) with the Golgi apparatus. J Cell Biol 95:405a (abstract)
Cataldo AM, Broadwell RD (1984) Cytochemical staining of the endoplasmic reticulum and glycogen in mouse anterior pituitary cells. J Histochem Cytochem 32:1285–1294
Chen JW, Murphy TL, Willingham MC, Pastan I, August JT (1985) Identification of two lysosomal membrane glycoproteins. J Cell Biol 101:85–95
Cheng H, Farquhar MG (1976) Presence of adenylate cyclase activity in Golgi and other fractions from rat liver. J Cell Biol 70:671–684
Chicheportiche Y, Vassalli P, Tartakoff AM (1984) Characterization of cytoplasmically oriented Golgi proteins with a monoclonal antibody. J Cell Biol 99:2200–2210
Cho M-I, Garant PR (1981a) Sequential events in the formation of collagen secretion granules with special reference to the development of segment-long-spacing-like aggregates. Anat Rec 199:309–320
Cho M-I, Garant PR (1981b) Role of microtubules in the organization of the Golgi complex and the secretion of collagen secretory granules by periodontal ligament fibroblasts. Anat Rec 199:459–471
Christensen NJ, Rubin CE, Cheung MC, Albers JJ (1983) Ultrastructural immunolocalization of apolipoprotein-B within human jejunal absorptive cells. J Lipid Res 24:1229–1242
Claude A (1970) Growth and differentiation of cytoplasmic membranes in the course of lipoprotein granule synthesis in the hepatic cell. I. Elaboration of elements of the Golgi complex. J Cell Biol 47:745–766
Clermont Y, Tang XM (1985) Glycoprotein synthesis in the Golgi apparatus of spermatids during spermiogenesis of the rat. Anat Rec 213:33–43
Clermont Y, Lalli M, Rambourg A (1981) Ultrastructural localization of nicotinamide adenine dinucleotide phosphatase (NADPase), thiamine pyrophosphatase (TPPase) and cytidine monophosphatase (CMPase) in the Golgi apparatus of early spermatids of the rat. Anat Rec 201:613–622
Colman A, Jones EA, Heasman J (1985) Meiotic maturation in *Xenopus* oocytes: A link between cessation of protein secretion and the polarized disappearance of Golgi apparati. J Cell Biol 101:313–318
Creek KE, Sly WS (1984) The role of the phosphomannosyl receptor in the transport of acid hydrolyses to lysosomes. In: Dingle JT, Dean RT, Sly WS (eds) Lysosomes in biology and pathology 7. Amsterdam, Elsevier, pp 63–82
Croze EM, Morré DJ, Morré DM, Kartenbeck JK, Franke WW (1982) Distribution of clathrin and spiny coated vesicles on membranes within mature Golgi apparatus elements of mouse liver. Eur J Cell Biol 28:130–137
Danielsen EM, Cowell GM, Poulsen SS (1983) Biosynthesis of intestinal microvillar proteins. Role of the Golgi complex and microtubules. Biochem J 216:37–42
Danielsen EM, Cowell GM, Hansen GH, Gorr S-U, Sjöström H, Noren O (1986) Role of the Golgi complex and characteristics of post-Golgi transport in the biosynthesis of intestinal microvillar proteins. Biochem Soc Trans 14:165–170
Debray H, Decout D, Strecker G, Spik G, Montreuil J (1981) Specificity of twelve lectins towards oligosaccharides and glycopeptides related to N-glycosyl proteins. Eur J Biochem 117:41–55
DeCamilli P, Moretti M, Donini SD, Walter U, Lohmann SM (1986) Heterogeneous distribution of the cAMP receptor protein RII in the nervous system: Evidence for its intracellular accumulation on microtubules, microtubule-organizing centers, and in the area of the Golgi complex. J Cell Biol 103:189–203
Decker RS (1974) Lysosomal packaging in differentiating and degenerating anuran lateral motor column neurons. J Cell Biol 61:599–612
Deschuyteneer M, Prieels J-P, Mosselmans R (1984) Galactose-specific adsorptive endocytosis: An ultrastructural qualitative and quantitative study in cultured rat hepatocytes. Biol Cell 50:17–30
Deutscher SL, Creek KE, Merion M, Hirschberg CB (1983) Subfractionation of rat liver Golgi apparatus: separation of enzyme activities involved in the biosynthesis of phosphomannosyl recognition marker in lysosomal enzymes. Proc Natl Acad Sci USA 80:3938–3942
Deutscher SL, Hirschberg CB (1986) Mechanism of galactosylation in the Golgi apparatus. A chinese hamster ovary cell mutant deficient in translocation of UDP-galactose across Golgi vesicle membranes. J Biol Chem 261:96–100
Doine AI, Oliver C, Hand AR (1984) The Golgi apparatus and GERL during postnatal

differentiation of rat parotid acinar cells: An electron microscopic cytochemical study. J Histochem Cytochem 32:477–485
Doty SB, Smith CE, Hand AR, Oliver C (1977) Inorganic trimetaphosphatase as a histochemical marker for lysosomes in light and electron microscopy. J Histochem Cytochem 25:1381–1384
Doyle C, Roth MG, Sambrook J, Gething M-J (1985) Mutations of the cytoplasmic domain of the influenza virus hemagglutinin affect different stages of intracellular transport. J Cell Biol 100:704–714
Dunphy WG, Rothman JE (1983) Compartmentation of asparagine-linked oligosaccharide processing in the Golgi apparatus. J Cell Biol 97:270–275
Dunphy WG, Rothman JE (1985) Compartmental organization of the Golgi stack. Review. Cell 42:13–21
Dunphy WG, Brands R, Rothman JE (1985) Attachment of terminal N-acetylglucosamine to asparagine-linked oligosaccharides occurs in central cisternae of the Golgi stack. Cell 40:463–472
Ede DA, Wilby OK (1981) Golgi orientation and cell behaviour in the developing pattern of chondrogenic condensation in chick limb-bud mesenchyme. Histochemistry 13:615–630
Elhammer A, Kornfeld S (1984) Two enzymes involved in the synthesis of O-linked oligosaccharides are located on membranes of different densities in mouse lymphoma BW 5147 cells. J Cell Biol 98:327–331
Ellinger A, Pavelka M (1982) The Golgi apparatus of rat small intestinal absorptive cells. II. Morphology and cytochemical staining pattern during cell differentiation. J Submicrosc Cytol 14:587–596
Ellinger A, Pavelka M (1984a) Effect of monensin on the Golgi apparatus of absorptive cells in the small intestine of the rat. Morphological and cytochemical studies. Cell Tiss Res 235:187–194
Ellinger A, Pavelka M (1984b) Colchicine-induced tubular, vesicular and cisternal organelle aggregates in absorptive cells of the small intestine of the rat. I. Morphology and phosphatase cytochemistry. Biol Cell 52:43–52
Ellinger A, Pavelka M (1985) Post-embedding localization of glycoconjugates by means of lectins on thin sections of tissues embedded in LR white. Histochem J 17:1321–1336
Ellinger A, Pavelka M (1986) Colchicine-induced tubular, vesicular and cisternal organelle aggregates in absorptive cells of the small intestine of the rat. II. Endocytosis studies. Biol Cell 58:31–42
Ellinger A, Pavelka M, Gangl A (1983) Effect of colchicine on the rat small intestinal absorptive cells. II. Distribution of label after incorporation of (3H) fucose into plasma membrane glycoproteins. J Ultrastruct Res 85:260–271
Fambrough DM, Devreotes PN (1978) Newly synthesized acetylcholin receptors are located in the Golgi apparatus. J Cell Biol 76:237–244
Farquhar MG (1978) Recovery of surface membrane in anterior pituitary cells. J Cell Biol 77:R35–R42
Farquhar MG (1982) Membrane recycling in secretory cells: Pathway to the Golgi complex. In: Evered D, Collins GM (eds) Membrane recycling. Pitman, London, pp 157–183 (Ciba Foundation Symposium 92)
Farquhar MG (1983) Multiple pathways of exocytosis, endocytosis, and membrane recycling: Validation of a Golgi route. Fed Proc 42:2407–2413
Farquhar MG (1985) Progress in unraveling pathways of the Golgi traffic. Ann Rev Cell Biol 1:447–488
Farquhar MG, Palade GE (1981) The Golgi apparatus (complex) – (1954–1981) – from artifact to center stage. J Cell Biol 91:77s–103s
Farquhar MG, Bergeron JJM, Palade GE (1974) Cytochemistry of Golgi fractions prepared from rat liver. J Cell Biol 60:8–25
Fatem SH, Leblond CP (1985) Sulfation and transport of basement membrane proteoglycans, as visualized by 35S-sulfate radioautography in the endodermal cells of the rat parietal yolk sac. Am J Anat 173:127–145
Featherstone C, Griffiths G, Warren G (1985) Newly synthesized G protein of vesicular stomatitis virus is not transported to the Golgi complex in mitotic cells. J Cell Biol 101:2036–2046
Feldmann G, Maurice M, Bernuau D, Rogier E (1985) Morphological aspects of plasma protein synthesis and secretion by the hepatic cells. Int Rev Cytol 96:157–188
Fitting T, Kabat D (1982) Evidence for a glycoprotein "signal" involved in transport between

subcellular organelles. Two membrane glycoproteins encoded by murine leukemia virus reach the cell surface at different rates. J Biol Chem 257:14011–14017
Fleischer B (1983) Mechanisms of glycosylation in the Golgi apparatus. J Histochem Cytochem 31:1033–1040
Flickinger CJ (1973) Maintenance and regeneration of cytoplasmic organelles in hybrid amoebae formed by nuclear transplantation. Exp Cell Res 80:31–46
Flickinger CJ (1978) The pattern of appearance of enzymic activity during the development of the Golgi apparatus in amoebae. J Cell Sci 34:53–63
Franke WW, Lüder MR, Kartenbeck J, Zerban H, Keenan TW (1976) Involvement of vesicle coat material in casein secretion and surface regeneration. J Cell Biol 69:173–195
Fransen JAM, Ginsel LA, Hauri H-P, Sterchi E, Blok J (1985) Immuno-electronmicroscopical localization of a microvillus membrane disaccharidase in the human small-intestinal epithelium with monoclonal antibodies. Eur J Cell Biol 38:6–15
Friedman HI, Cardell RR Jr (1976) Alterations in the endoplasmic reticulum and Golgi complex of intestinal epithelial cells during fat absorption and after termination of this process: A morphological and morphometric study. Anat Rec 188:77–102
Friend DS (1969) Cytochemical staining of multivesicular body and Golgi vesicles. J Cell Biol 41:269–279
Friend DS, Farquhar MG (1967) Functions of coated vesicles during protein absorption in the rat vas deferens. J Cell Biol 35:357–376
Friend DS, Murray MJ (1965) Osmium impregnation of the Golgi apparatus. Am J Anat 117:135–150
Fries E, Gustafsson L, Peterson PA (1984) Four secretory proteins synthesized by hepatocytes are transported from endoplasmic reticulum to Golgi complex at different rates. EMBO J 3:147–152
Fritzler MJ, Etherington J, Sokoluk C, Kinsella TG, Valencia DW (1984) Antibodies from patients with autoimmune disease react with a cytoplasmic antigen in the Golgi apparatus. J Immunol 132:2904–2908
Fujita H, Sawano F (1983) On the internal polarity of the Golgi apparatus with special regard to its relationship to GERL. J Histochem Cytochem 31:227–229
Fuller SD, Bravo R, Simons K (1985) An enzymatic assay reveals that proteins destined for the apical and basolateral domains of an epithelial cell line share the same late Golgi compartments. EMBO J 4:297–307
Gabel CA, Bergmann JE (1985) Processing of the asparagine-linked oligosaccharides of secreted and intracellular forms of the vesicular stomatitis virus G protein: in vivo evidence of Golgi apparatus compartmentalization. J Cell Biol 101:460–469
Gahmberg N, Kuismanen E, Keränen S, Pettersson RF (1986) Uukuniemi virus glycoproteins accumulate in and cause morphological changes of the Golgi complex in the absence of virus maturation. J Virol 57:899–906
Garcia-Porrero JA, Icardo JM, Ojeda JL (1981) A quantitative study of the position of the Golgi apparatus in the early developing chick eye. Anat Embryol 163:77–85
Geisow MJ (1982) Intracellular membrane traffic. Nature 295:649–650
Geuze JJ, Kramer M (1974) Function of coated membranes and multivesiculated bodies during membrane regulation in stimulated exocrine pancreatic cells. Cell Tiss Res 156:1–20
Geuze HJ, Slot JW (1980) The subcellular localization of immunoglobulin in mouse plasma cells, as studied with immunoferritin cytochemistry on ultrathin frozen sections. Am J Anat 158:161–169
Geuze JJ, Slot JW, Tukuyasu KT, Goedemans WEM, Griffith JM (1979) Immunocytochemical localization of amylase and chymotrypsinogen in the exocrine pancreatic cell with special attention to the Golgi complex. J Cell Biol 82:697–707
Geuze HJ, Slot JW, Strous GJAM, Lodish HF, Schwartz AL (1982) Immunocytochemical localization of the receptor for asialoglycoprotein in rat liver cells. J Biol 92:865–870
Geuze HJ, Slot JW, Strous GJAM, Lodish HF, Schwartz AL (1983a) Intracellular site of asialoglycoprotein receptor-ligand uncoupling: Double-label immunoelectron microscopy during receptor-mediated endocytosis. Cell 32:277–287
Geuze HJ, Slot JW, Strous GJAM, Schwartz AL (1983b) The pathway of the asialoglycoprotein-ligand during receptor-mediated endocytosis: A morphological study with colloidal gold ligand in the human hepatoma cell line, Hep G2. Eur J Cell Biol 32:38–44
Geuze HJ, Slot JW, Strous GJAM, Hasilik A, von Figura K (1984a) Ultrastructural localization of the mannose-6-phosphate receptor in rat liver. J Cell Biol 98:2047–2054

Geuze HJ, Slot JW, Strous GJAM, Peppard J, von Figura K, Hasilik A, Schwartz AL (1984b) Intracellular receptor sorting during endocytosis: Comparative immunoelectron microscopy of multiple receptors in rat liver. Cell 37:195–204

Geuze HJ, Slot JW, Strous GJ, Luzio JP, Schwartz AL (1984c) A cycloheximide-resistant pool of receptors for asialoglycoproteins and mannose-6-phosphate residues in the Golgi complex of hepatocytes. EMBO J 3:2677–2685

Geuze HJ, Slot JW, Strous GJAM, Hasilik A, von Figura K (1985) Possible pathways for lysosomal enzyme delivery. J Cell Biol 101:2253–2262

Glickman RM, Perrotto JL, Kirsch K (1976) Intestinal lipoprotein formation: Effects of colchicine. Gastroenterology 70:347–352

Glickman J, Croen K, Kelly S, Al-Awqati Q (1983) Golgi membranes contain an electrogenic H^+-pump in parallel to a chloride conductance. J Cell Biol 97:1303–1308

Goldberg DE, Kornfeld S (1983) Evidence for extensive subcellular organization of asparagine-linked oligosaccharide processing and lysosomal enzyme phosphorylation. J Biol Chem 258:3159–3165

Goldenberg R, Fine RE (1984) Coated vesicles purified from chick tendon fibroblasts contain newly synthesized type I procollagen. Exp Cell Res 157:41–49

Goldfischer S (1982) The internal reticular apparatus of Camillo Golgi: A complex heterogenous organelle, enriched in acid, neutral, and alkaline phosphatases, and involved in glycosylation, secretion, membrane flow, lysosome formation, and intracellular digestion. J Histochem Cytochem 30:717–733

Goldfischer S, Essner E, Schiller B (1971) Nucleoside diphosphatase and thiamine pyrophosphatase activities in the endoplasmic reticulum and Golgi apparatus. J Histochem Cytochem 19:349–360

Goldstein IJ, Hayes CE (1978) The lectins: Carbohydrate-binding proteins of plants and animals. Adv Carbohydr Chem Biochem 35:127–340

Golgi C (1898) Sur la structure des cellules nerveuses. Arch Ital Biol 30:60–71

Gonatas NK, Kim SU, Stieber A, Avrameas S (1977) Internalization of lectins in neuronal GERL. J Cell Biol 73:1–13

Gonatas NK, Stieber A, Hickey WF, Herbert SH, Gonatas JO (1984) Endosomes and Golgi vesicles in adsorptive and fluid phase endocytosis. J Cell Biol 99:1379–1390

Gorelick FS, Sarras MP Jr, Jamieson JD (1982) Regional differences in the lectin binding to colonic epithelium by fluorescence and electron microscopy. J Histochem Cytochem 30:1097–1108

Gorvel J-P, Massey D, Maroux S (1986) Evidence for selective transport of two brush-border glycoproteins from endoplasmic reticulum to Golgi complex in rabbit enterocytes. Biol Cell 56:251–254

Green J, Griffiths G, Louvard D, Quinn P, Warren G (1980) Passage of viral membrane proteins through the Golgi complex. J Mol Biol 152:663–698

Griffiths G, Brands R, Burke B, Louvard D, Warren G (1982) Viral membrane proteins acquire galactose in trans Golgi cisternae during intracellular transport. J Cell Biol 95:781–792

Griffiths G, Quinn P, Warren G (1983) Dissection of the Golgi complex. I. Monensin inhibits the transport of viral membrane proteins from medial to trans Golgi cisternae in baby hamster kidney cells infected with Semliki forest virus. J Cell Biol 96:835–850

Griffiths G, Pfeiffer S, Simons K, Matlin K (1985) Exit of newly synthesized membrane proteins from the trans cisterna of the Golgi complex to the plasma membrane. J Cell Biol 101:949–964

Grove SN, Bracker CE, Morré DJ (1968) Cytomembrane differentiation in the endoplasmic reticulum-Golgi apparatus-vesicle complex. Science 161:171–173

Guan J-L, Machamer CE, Rose JK (1985) Glycosylation allows cell-surface transport of an anchored secretory protein. Cell 42:489–496

Guillouzo A, Beaumont C, Le Rumeur E, Rissel M, Latinier M-F, Guguen-Guillouzo C, Bourel M (1982) New findings on immunolocalization of albumin in rat hepatocytes. Biol Cell 43:163–172

Haimes HB, Stockert RJ, Morell AC, Novikoff AB (1981) Carbohydrate-specified endocytosis: Colocalization of ligand in the lysosomal compartment. Proc Natl Acad Sci USA 78:6936–6939

Hand AR (1971) Morphology and cytochemistry of the Golgi apparatus of rat salivary gland acinar cells. Am J Anat 130:141–158

Hand AR (1980) Cytochemical differentiation of the Golgi apparatus from GERL. J Histochem Cytochem 28:82–86

Hand AR, Oliver C (1977a) Cytochemical studies of GERL and its role in secretory granule formation in exocrine cells. Histochem J 9:375–392

Hand AR, Oliver C (1977b) Relationship between the Golgi apparatus, GERL, and secretory granules in acinar cells of the rat exorbital lacrimal gland. J Cell Biol 74:399–413

Hand AR, Oliver C (1981) The Golgi apparatus: Protein transport and packaging in secretory cells. In: Hand AR, Oliver C (eds) Methods in cell biology, vol 23, Part 2. Academic, New York, pp 137–153

Hand AR, Oliver C (1984a) The role of GERL in the secretory process. In: Cantin M (ed) Cell biology of the secretory process. Karger, Basel, pp 148–170

Hand AR, Oliver C (1984b) Effects of secretory stimulation on the Golgi apparatus and GERL of rat parotid acinar cells. J Histochem Cytochem 32:403–412

Hanover JA, Willingham MC, Pastan I (1984) Kinetics of transit of transferrin and epidermal growth factor through clathrin-coated membranes. Cell 39:283–293

Harris N, Oparka KJ (1983) Connection between dictyosomes, ER and GERL in cotyledons of mung bean (*Vigna radiata* L.). Protoplasma 114:93–102

Hasilik A, von Figura K (1981) Oligosaccharides in lysosomal enzymes. Distribution of high-mannose and complex oligosaccharides in cathepsin D and beta-hexosaminidase. Eur J Biochem 121:125–129

Hauri H-P, Sterchi E, Bienz D, Fransen JAM, Marxer A (1985) Expression and intracellular transport of microvillus membrane hydrolases in human intestinal epithelial cells. J Cell Biol 101:838–851

Hedman K, Pastan I, Willingham MC (1986) The organelles of the trans domain of the cell. Ultrastructural localization of sialoglycoconjugates using Limax flavus agglutinin. J Histochem Cytochem 34:1069–1077

Helenius A, Mellman I, Wall D, Hubbard A (1983) Endosomes. Trends Biochem Sci 7:245–250

Hermo L, Clermont Y, Rambourg A (1979) Endoplasmic reticulum – Golgi apparatus relationships in the rat spermatid. Anat Rec 193:243–256

Hermo L, Rambourg A, Clermont Y (1980) Three-dimensional architecture of the cortical region of the Golgi apparatus in rat spermatids. Am J Anat 157:357–373

Herzog V (1983) Transcytosis in thyroid follicle cells. J Cell Biol 97:607–617

Herzog V (1984) Pathways of endocytosis in thyroid follicle cells. Int Rev Cytol 91:107–139

Herzog V (1985) Secretion of sulfated thyroglobulin. Eur J Cell Biol 39:399–409

Herzog V, Farquhar M (1977) Luminal membrane retrieved after exocytosis reaches most Golgi cisternae in secretory cells. Proc Natl Acad Sci USA 74:5073–5077

Herzog V, Reggio H (1980) Pathways of membrane retrieved from the luminal surface of exocrine cells of the pancreas. Eur J Cell Biol 21:141–150

Hickey WF, Stieber A, Hogue-Angeletti R, Gonatas J, Gonatas NK (1983) Nerve growth factor induced changes in the Golgi apparatus of PC-12 rat pheochromocytoma cells as studied by ligand endocytosis, cytochemical and morphometric methods. J Neurocytol 12:751–766

Higgins JA, Barrnett RJ (1971) Fine structural localization of acyltransferases. The monoglyceride and α-glycerophosphate pathways in intestinal absorptive cells. J Cell Biol 50:102–120

Higgins JA, Hutson JL (1984) The roles of Golgi and endoplasmic reticulum in the synthesis and assembly of lipoprotein lipids in rat hepatocytes. J Lipid Res 25:1295–1305

Hiller G, Weber K (1982) Golgi detection in mitotic and interphase cells by antibodies to secreted galactosyltransferase. Exp Cell Res 142:85–94

Hoflack B, Kornfeld S (1985) Lysosomal enzyme binding to mouse P 388D1 macrophage membranes lacking the 215-kDa mannose 6-phosphate receptor: Evidence for the existence of a second mannose 6-phosphate receptor. Proc Natl Acad Sci USA 82:4428–4432

Holtzman E, Novikoff AB, Villaverde H (1967) Lysosomes and GERL in normal and chromatolytic neurons of the rat ganglion nodosum. J Cell Biol 33:419–435

Hopkins C (1985) Coated pits and their role in membrane receptor internalization. In: Cohen H, Houslay MD (eds) Molecular mechanisms in transmembrane signalling. Elsevier, Amsterdam, New York, Oxford, pp 338–357

Howell KE, Palade GE (1982) Heterogeneity of lipoprotein particles in hepatic Golgi fractions. J Cell Biol 92:833–845

Hubbard SC, Ivatt RJ (1981) Synthesis and processing of asparagine-linked oligosaccharides. Ann Rev Biochem 50:555–583

Ichikawa M, Ichikawa A, Tanabe T (1982) High resolution analysis of three dimensional structure of the Golgi apparatus in rapid frozen, substitution fixed gerbil sublingual gland acinar cells. J Electr Microsc 31:397–401

Irimura T, Nicolson GL (1983) Carbohydrate-chain analysis by lectin binding to mixtures of glycoproteins separated by polyacrylamide slab-gel electrophoresis with in situ chemical modifications. Carbohydr Res 115:209–220

Ito A, Palade GE (1978) Presence of NADPH-cytochrome P-450 reductase in rat liver Golgi membranes. Evidence obtained by immunoadsorption method. J Cell Biol 79:590–597

Jaeken L, Thines-Sempoux D (1981) A three-dimensional study of organelle interrelationships in regenerating rat liver. 6. Golgi apparatus. Cell Biol Int Rep 5:261–273

Jamieson JD, Palade GE (1967a) Intracellular transport of secretory proteins in the pancreatic exocrine cell. I. Role of the peripheral elements of the Golgi complex. J Cell Biol 34:577–596

Jamieson JD, Palade GE (1967b) Intracellular transport of secretory proteins in the pancreatic exocrine cell. II. Transport to condensing vacuoles and zymogen granules. J Cell Biol 34:597–615

Jamieson JD, Palade GE (1968) Intracellular transport of secretory proteins in the pancreatic exocrine cell. IV. Metabolic requirements. J Cell Biol 39:589–603

Jamieson JD, Palade GE (1971) Synthesis, intracellular transport, and discharge of secretory proteins in stimulated pancreatic exocrine cells. J Cell Biol 50:135–158

Jersild RA Jr (1966) A time sequence study of fat absorption in the rat jejunum. Am J Anat 118:135–162

Jones AL, Ockner RK (1971) An electron microscopic study of endogenous very low density lipoprotein production in the intestine of rat and man. J Lipid Res 12:580–589

Joseph KC, Stieber A, Gonatas NK (1979) Endocytosis of cholera toxin in GERL-like structures of murine neuroblastoma cells pretreated with GM1 ganglioside. J Cell Biol 81:543–554

Kaplan F, Hechtman P (1984) Rat liver Golgi galactosyltransferases. Distinct enzymes for glycolipid and glycoprotein acceptor substrates. Biochem J 217:353–364

Karim A, Cournil I, Leblond CP (1979) Immunocytochemical localization of procollagens. II. Electron microscopic distribution of procollagen I antigenicity in the odontoblasts and predentin of rat incisor teeth by a direct method using horseradish peroxidase-linked antibodies. J Histochem Cytochem 27:1070–1083

Kartenbeck J, Schmid E, Müller H, Franke WW (1981) Immunological identification and localization of clathrin and coated vesicles in cultured cells and in tissues. Exp Cell Res 133:191–211

Kay DG, Khan MN, Posner BI, Bergeron JJM (1984) In vivo uptake of insulin into hepatic Golgi fractions: Application of the diaminobenzidine-shift protocol. Biochem. Biophys Res Commun 123:1144–1148

Kayahara T (1982) The fine localization of acid phosphatase activity in the unvacuolated notochordal cells of the early chick embryo. Histochem J 14:347–360

Keenan TW, Morré DJ, Basu S (1974) Ganglioside biosynthesis: Concentration of glycerosphingolipid glycosyltransferases in Golgi apparatus from rat liver. J Biol Chem 249:310–315

Kelly RB (1985) Pathways of protein secretion in eukaryotes. Science 230:25–32

Kessel RG (1971) Origin of the Golgi apparatus in eukaryotic cells of the grasshopper. J Ultrastruct Res 34:260–275

Kimura M, Ichihara I (1985) Cytochemical studies of the acid phosphatases in the rat lateral prostate with special reference to secretory apparatus, and lysosomal system. Histochemistry 82:519–523

Kimura JH, Lohmander LS, Hascall VC (1984) Studies on the biosynthesis of cartilage proteoglycan in a model system of cultured chondrocytes from the swarm rat chondrosarcoma. J Cell Biochem 26:261–278

Knudson CM, Stemberger BH, Patton S (1978) Effects of colchicine on ultrastructure of the lactating mammary cell: Membrane involvement and stress on the Golgi apparatus. Cell Tiss Res 195:169–181

Kornfeld K, Reitman ML, Kornfeld R (1981) The carbohydrate-binding specificity of pea and lentil lectins. Fucose is an important determinant. J Biol Chem 256:6633–6640

Kornfeld R, Kornfeld S (1985) Assembly of asparagine-linked oligosaccharides. Ann Rev Biochem 54:631–664

Kraehenbuhl JP, Racine L, Jamieson JD (1977) Immunocytochemical localization of secretory proteins in bovine pancreatic exocrine cells. J Cell Biol 72:406–423

Kuhn NJ, White A (1977) The role of nucleoside diphosphatase in an uridine nucleotide cycle associated with lactose synthesis in rat mammary gland Golgi apparatus. Biochem J 168:423–433

Kupfer A, Louvard D, Singer SJ (1982) Polarization of the Golgi apparatus and the microtubule-organizing center in cultured fibroblasts at the edge of an experimental wound. Proc Natl Acad Sci USA 79:2603–2607

Kupfer A, Dennert G, Singer SJ (1983) The reorientation of the Golgi apparatus and the microtubule-organizing center in the cytotoxic effector cell is a prerequisite in the lysis of bound target cells. J Mol Cell Immunol 2:37–49

Kurosumi K (1984) Ultrastructure and function of Golgi apparatus as revealed by electronmicroscopic cytochemistry and immunocytochemistry. In: Seno S, Okada Y (eds) International Cell Biology 1984, Academic, Tokyo, p 96

Laurie GW, Leblond CP, Martin GR (1982a) Intracellular localization of basement membrane precursors in the endodermal cells of the rat parietal yolk sac. II. Immunostaining for type IV collagen and its precursors. J Histochem Cytochem 30:983–990

Laurie GW, Leblond CP, Martin GR, Silver MH (1982b) Intracellular localization of basement membrane precursors in the endodermal cells of the rat parietal yolk sac. III. Immunostaining for laminin and its precursors. J Histochem Cytochem 30:991–998

Lee RWH, Huttner WB (1985) (Glu62, Ala36, Tyr8)n serves as high-affinity substrate for tyrosylprotein sulfotransferase: A Golgi enzyme. Proc Natl Acad Sci USA 82:6143–6147

Levine JS, Allen RH, Alpers DH, Seetharam B (1984) Immunocytochemical localization of the intrinsic factor-cobalamine receptor in dog ileum: Distribution of intracellular receptor during cell maturation. J Cell Biol 98:1111–1118

Lin JJ-C, Queally SA (1982) A monoclonal antibody that recognizes Golgi-associated protein of cultured fibroblasts cells. J Cell Biol 92:108–112

Lipsky NG, Pagano RE (1985a) A vital stain for the Golgi apparatus. Science 228:745–747

Lipsky NG, Pagano RE (1985b) Intracellular translocation of fluorescent sphingolipids in cultured fibroblasts: Endogenously synthesized sphingomyelin and glucocerebroside analogues pass through the Golgi apparatus en route to the plasma membrane. J Cell Biol 100:27–34

Locke M, Huie P (1976a) The beads in the Golgi complex/endoplasmic reticulum region. J Cell Biol 70:384–394

Locke M, Huie P (1976b) Vertebrate Golgi complexes have beads in a similar position to those found in arthropods. Tiss Cell 8:739–743

Locke M, Huie P (1983) The mystery of the unstained Golgi complex cisternae. J Histochem Cytochem 31:1019–1032

Lodish HF, Kong N, Snider M, Strous GJAM (1983) Hepatoma secretory proteins migrate from rough endoplasmic reticulum to Golgi at characteristic rates. Nature 304:80–83

Lucocq J, Montesano R (1985) Nonrandom positioning of Golgi apparatus in pancreatic B cells. Anat Rec 213:182–186

Lucocq JM, Brada D, Roth J (1986) Immunolocalization of the oligosaccharide trimming enzyme glucosidase II. J Cell Biol 102:2137–2146

Louvard D, Reggio H, Warren G (1982) Antibodies to the Golgi complex and rough endoplasmic reticulum. J Cell Biol 92:92–107

Magalhaes MC, Vitor AB, Magalhaes MM (1985) Effects of vinblastine and colchicine on the rat adrenal cortex: Morphometric and cytochemical studies. J Ultrastr Res 91:149–158

Matlin KS, Simons K (1983) Reduced temperature prevents transfer of a membrane glycoprotein to the cell surface but does not prevent terminal glycosylation. Cell 34:233–243

Matsuura S, Tashiro Y (1979) Immunoelectronmicroscopic studies of endoplasmic reticulum – Golgi relationships in the intracellular transport process of lipoprotein particles in rat hepatocytes. J Cell Sci 39:273–290

Matyas GR, Morré DJ (1983) Coupling of uridine 5′-diphosphate (UDP) formation and nicotinamide adenine dinucleotide (NAD^+) reduction for cytochemical localization of glycosyltransferases. J Histochem Cytochem 31:1175–1182

McClintock J, Locke M (1982) Lead staining in the Golgi complex. Tiss Cell 14:541–554

McFadden GI, Melkonian M (1986) Golgi apparatus activity and membrane flow during

scale biogenesis in the green flagellate *Scherffelie dubia* (*Prasinophyceae*): I. Flagellar regeneration. Protoplasma 130:186–198
Melmed RN, Benitez CJ, Holt SJ (1973) An ultrastructural study of the pancreatic acinar cell in mitosis, with special reference to changes in the Golgi complex. J Cell Sci 12:163–173
Merisco EM, Farquhar MG, Palade GE (1982) Coated vesicle isolation by immunoadsorption on *Staphylococcus aureus* cells. J Cell Biol 92:846–858
Michaels JE (1983) The effects of colchicine on the distribution of glycoprotein-containing vesicles in epithelial cells of murine colon. Cell Tiss Res 228:323–335
Miller JS, Gavino VC, Ackerman GA, Sharma HM, Milo GE, Geer JC, Cornwell DG (1980) Triglycerides, lipid droplets, and lysosomes in aorta smooth muscle cells during the control of cell proliferation with polyunsaturated fatty acids and vitamin E. Lab Invest 42:495–507
Mizuno M, Brown WR, Vierling JM (1984) Ultrastructural immunocytochemical localization of the asialoglycoprotein receptor in rat hepatocytes. Gastroenterology 87:763–769
Mollenhauer HH (1965) An intercisternal structure in the Golgi apparatus. J Cell Biol 24:504–511
Mollenhauer HH, Morré DJ (1976) Transition elements between endoplasmic reticulum and Golgi apparatus in plant cells. Cytobiologie 13:297–306
Mollenhauer HH, Hass BS, Morré DJ (1976) Membrane transformation in Golgi apparatus of rat spermatids. A role for thick cisternae and two classes of coated vesicles in acrosome formation. J Microsc Biol Cell 27:33–36
Montreuil J (1984) Spatial conformation of glycans and glycoproteins. Biol Cell 51:115–132
Morré DJ, Ovtracht L (1977) Dynamics of the Golgi apparatus: Membrane differentiation and membrane flow. Int Rev Cytol [Suppl] 5:61–188
Morré DJ, Ovtracht L (1981) Structure of rat liver Golgi apparatus: Relationship to lipoprotein secretion. J Ultrastruct Res 74:284–295
Morré DJ, Mollenhauer HH (1983) Dictyosome polarity and membrane differentiation in outer cap cells of the maize root tip. Eur J Cell Biol 29:126–132
Morré DJ, Vigil EL, Frantz C, Goldenberg H, Crane FL (1978) Cytochemical demonstration of the glutaraldehyde-resistant NADH ferricyanide oxido-reductase activities in rat liver plasma membranes and Golgi apparatus. Cytobiologie 18:213–230
Morré DJ, Kartenbeck J, Franke WW (1979) Membrane flow and interconversions among endomembranes. Biochim Biophys Acta 559:71–152
Morré DJ, Morré DM, Heidrich H-G (1983) Subfractionation of the rat liver Golgi apparatus by free-flow electrophoresis. Eur J Cell Biol 31:263–274
Morré DJ, Widnell CJ, Thilo L (1984a) Membrane dynamics: Flow routes and quantitation of membrane transport and recycling. Fed Proc 43:2884–2887
Morré DJ, Creek KE, Matyas GR, Minnifield N, Sun I, Baudoin P, Morré DM, Crane FL (1984b) Free-flow electrophoresis for subfractionation of rat liver Golgi apparatus. Biotechniques 2:224–233
Nemere I, Kupfer A, Singer SJ (1985) Reorientation of the Golgi apparatus and the microtubule-organzing center inside macrophages subjected to a chemotactic gradient. Cell Motility 5:17–29
Neutra M, Leblond CP (1966) Synthesis of the carbohydrate mucus in the Golgi complex as shown by electron microscope radioautography of goblet cells from rats injected with glucose-H3. J Cell Biol 30:119–136
Newman GR, Jasani B, Williams D (1983) A simple post-embedding system for the rapid demonstration of tissue antigens under the electron microscope. Histochem J 15:543–555
Nigg EA, Schäfer G, Hilz H, Eppenberger HM (1985a) Cyclic-AMP-dependent proteinkinase type II is associated with the Golgi complex and with centrosomes. Cell 41:1039–1051
Nigg EA, Hilz H, Eppenberger HM, Dutly F (1985b) Rapid and reversible translocation of the catalytic subunit of cAMP-dependent protein kinase type II from Golgi complex to the nucleus. EMBO J 4:2801–2806
Noda T, Ogawa K (1984) Golgi apparatus is one continuous organelle in pancreatic exocrine cell of the mouse. Acta Histochem Cytochem 17:435–451
Northcote DH (1979) The involvement of the Golgi apparatus in the biosynthesis and secretion of glycoproteins and polysaccharides. In: Manson LA (ed) Biomembranes, vol 10. Plenum, New York, pp 51–76
Novick P, Ferro S, Schekman R (1981) Order of events in the yeast secretory pathway. Cell 25:461–469

Novikoff AB (1964) GERL, its form and function in neurons of rat spinal ganglia. Biol Bull 127:358
Novikoff AB (1976) The endoplasmic reticulum: A cytochemist's view. (A Review) Proc Natl Acad Sci USA 73:2781–2787
Novikoff AB, Goldfischer S (1961) Nucleoside diphosphatase activity in the Golgi apparatus and its usefulness for cytological studies. Proc Natl Acad Sci USA 47:802–810
Novikoff AB, Novikoff PM (1977) Cytochemical contribution of differentiating GERL from the Golgi apparatus. Histochem J 9:525–551
Novikoff AB, Albana A, Biempica L (1968) Ultrastructural and cytochemical observations on B-16 and Harding-Passey mouse melanomas. The origin of premelanosomes and compound melanosomes. J Histochem Cytochem 16:299–319
Novikoff AB, Yam A, Novikoff PM (1975) Cytochemical study of secretory process in transplantable insulinoma of Syrian golden hamster. Proc Natl Acad Sci USA 72:4501–4505
Novikoff AB, Mori M, Quintana N, Yam A (1977) Studies on the secretory process in the mammalian exocrine pancreas. I. The condensing vacuoles. J Cell Biol 75:148–165
Novikoff PM, Yam A (1978) The cytochemical demonstration of GERL in rat hepatocytes during lipoprotein mobilisation. J Histochem Cytochem 26:1–13
Novikoff PM, Novikoff AB, Quintana N, Hauw JJ (1971) Golgi apparatus, GERL, and lysosomes in neurons in rat dorsal root ganglia, studied by thick section and thin section cytochemistry. J Cell Biol 50:859–886
Novikoff PM, Yam A, Novikoff AB (1981) Lysosomal compartment of macrophages: Extending the definition of GERL. Proc Natl Acad Sci USA 78:5699–5703
Novikoff PM, La Russo NF, Novikoff AB, Stockert RJ, Yam A, Le Sage GD (1983a) Immunocytochemical localization of lysosomal β-galactosidase in rat liver. J Cell Biol 97:1559–1565
Novikoff PM, Tulsiani DRP, Touster O, Yam A, Novikoff AB (1983b) Immunocytochemical localization of α-D-mannosidase II in the Golgi apparatus of rat liver. Proc Natl Acad Sci USA 80:4364–4368
Oliver C (1980) Cytochemical localization of acid phosphatase and trimetaphosphatase activities in exocrine acinar cells. J Histochem Cytochem 28:78–81
Oliver C (1983) Characterization of basal lysosomes in exocrine acinar cells. J Histochem Cytochem 31:1209–1216
Oliver C, Hand AR (1983) Enzyme modulation of the Golgi apparatus and GERL: A cytochemical study of parotid acinar cells. J Histochem Cytochem 31:1041–1048
Oliver C, Auth RE, Hand AR (1980) Morphological and cytochemical alterations of the Golgi apparatus and GERL in rat parotid acinar cells during ethionine intoxication and recovery. Am J Anat 158:275–284
Oliver JM, Berlin RD (1982) Mechanisms that regulate the structural and functional architecture of the cell surfaces. Int Rev Cytol 74:55–94
Ono K (1979) Ultrastructural localization of acid phosphatase activity in the small intestinal absorptive cells of adult rats. Histochemistry 62:113–124
Oomori Y, Ono K, Ishikawa K, Satoh Y-I, Matoba M (1984) Ultrastructural localization of thiamine pyrophosphatase activity in the intestinal epithelial cells of adult rats. Acta Histochem 74:181–187
Orci L (1982) Macro- and micro-domains in the endocrine pancreas. Diabetes 31:538–565 (Banting Lecture 1981)
Orci L, Montesano R, Meda P, Malaisse-Lagae F, Brown D, Perrelet A, Vassalli P (1981) Heterogenous distribution of filipin-cholesterol complexes across the cisternae of the Golgi apparatus. Proc Natl Acad Sci USA 78:293–297
Orci L, Ravazzola M, Perrelet A (1984a) (Pro)insulin associates with Golgi membranes of pancreatic B cells. Proc Natl Acad Sci USA 81:6743–6746
Orci L, Halban P, Amherdt R, Ravazzola M, Vassalli J-D, Perrelet A (1984b) A clathrin-coated, Golgi-related compartment of the insulin secreting cell accumulates proinsulin in the presence of monensin. Cell 39:39–47
Orci L, Halban P, Amherdt M, Ravazzola M, Vassalli J-D, Perrelet A (1984c) Non-converted, aminoacid analog-modified proinsulin stays in a Golgi-derived clathrin-coated membrane compartment. J Cell Biol 99:2187–2192
Orci L, Ravazzola M, Amherdt M, Louvard D, Perrelet A (1985a) Clathrin-immunoreactive sites in the Golgi apparatus are concentrated at the trans pole in polypeptide hormone-secreting cells. Proc Natl Acad Sci USA 82:5385–5389

Orci L, Ravazzola M, Amherdt M, Madsen O, Vassalli J-D, Perellet A (1985b) Direct identification of prohormone conversion site in insulin-secreting cells. Cell 42:671–681

Orci L, Glick BS, Rothman JE (1986) A new type of coated vesicular carrier that appears not to contain clathrin: Its possible role in protein transport within the Golgi stack. Cell 46:171–184

Ottosen PD, Courtoy PJ, Farquhar MG (1980) Pathways followed by membrane recovered from the surface of plasma cells and myeloma cells. J Exp Med 152:1–19

Ovtracht L, Thiéry J-P (1972) Mise en évidence par cytochimie ultrastructurale de compartiments physiologiquement différents dans un même saccule Golgien. J Microsc 15:135–170

Ovtracht L, Morré DJ, Cheetham RD, Mollenhauer HH (1973) Subfractionation of Golgi apparatus from rat liver: method and morphology. J Microsc 18:87–102

Paavola L (1978a) The corpus luteum of the guinea pig. II. Cytochemical studies on the Golgi complex, GERL, and lysosomes in luteal cells during maximal progesteron secretion. J Cell Biol 79:45–58

Paavola L (1978b) The corpus luteum of the guinea pig. III. Cytochemical studies on the Golgi complex and GERL during normal postpartum regression of luteal cells, emphasizing the origin of lysosomes and autophagic vacuoles. J Cell Biol 79:59–73

Paavola LG, Strauss JF, Boyd CO, Nestler JE (1985) Uptake of gold- and 3H-cholesteryl linoleate-labeled human low density lipoprotein by cultured rat granulosa cells: Cellular mechanisms involved in lipoprotein metabolism and their importance to steroidogenesis. J Cell Biol 100:1235–1247

Palade GE (1975) Intracellular aspects of protein secretion. Science 189:347–358

Palade GE (1983) Membrane biogenesis: An overview. In: Fleischer S, Fleischer B (eds) Methods in enzymology – Biomembranes, part J, vol 96. Academic, New York, pp XXIX–LV

Pâquet MR, Pfeffer SR, Burczak JD, Glick BS, Rothman JE (1986) Components responsible for transport between successive Golgi cisternae are highly conserved in evolution. J Biol Chem 261:4367–4370

Parsons SM, Smith CE (1984) Ultrastructural localization of nicotinamide adenine dinucleotide phosphatase (NADPase) activity within columnar, goblet, and Paneth cells of the rat small intestine. J Histochem Cytochem 32:989–997

Pastan IH, Willingham MC (1981) Journey to the center of the cell: Role of the receptosome. Science 214:504–509

Pastan I, Willingham MC (1985) The pathways of endocytosis. In: Pastan I, Willingham M (eds) Endocytosis. Plenum, New York, pp 1–44

Patzelt C, Brown D, Jeanrenaud B (1977) Inhibitory effect of colchicine on amylase secretion by rat parotid glands. Possible localization in the Golgi area. J Cell Biol 73:578–593

Pavelka M, Ellinger A (1981a) Morphological and cytochemical studies on the Golgi apparatus of rat jejunal absorptive cells. J Ultrastruct Res 77:210–222

Pavelka M, Ellinger A (1981b) Effect of colchicine on the Golgi apparatus and on GERL of rat jejunal absorptive cells. Ultrastructural localization of thiamine pyrophosphatase and acid phosphatase activity. Eur J Cell Biol 24:53–61

Pavelka M, Ellinger A (1982) The Golgi apparatus of rat small intestinal absorptive cells. I. Morphology and cytochemical staining pattern in proximal and distal small intestinal regions. J Submicrosc Cytol 14:577–585

Pavelka M, Ellinger A (1983a) Effect of colchicine on the Golgi complex of rat pancreatic acinar cells. J Cell Biol 97:737–748

Pavelka M, Ellinger A (1983b) The trans Golgi face in rat small intestinal absorptive cells. Eur J Cell Biol 29:253–261

Pavelka M, Ellinger A (1985) Localization of binding sites for concanavalin A, Ricinus communis I and Helix pomatia lectin in the Golgi apparatus of rat small intestinal absorptive cells. J Histochem Cytochem 33:905–914

Pavelka M, Ellinger A (1986a) The Golgi apparatus in the acinar cells of the developing embryonic pancreas. I. Morphology and enzyme cytochemistry. Am J Anat 178:215–223

Pavelka M, Ellinger A (1986b) The Golgi apparatus in the acinar cells of the developing embryonic pancreas. II. Localization of lectin binding sites. Am J Anat 178:224–230

Pavelka M, Ellinger A (1986c) RCA I binding patterns of the Golgi apparatus. Eur J Cell Biol 41:270–278

Pavelka M, Ellinger A (1986d) Lectin-binding patterns of the Golgi apparatus. In: Bog-Hansen TC, van Driessche E (eds) Lectins, vol 5. De Gruyter, Berlin, pp 485–492

Pavelka M, Gangl A (1983) Effects of colchicine on the intestinal transport of endogenous lipid. Ultrastructural, biochemical, and radiochemical studies in fasting rats. Gastroenterology 84:544–555

Pavelka M, Ellinger A, Gangl A (1983) Effect of colchicine on rat intestinal absorptive cells. I. Formation of basolateral microvillus borders. J Ultrastruct Res 85:249–259

Pearse BMF (1976) Clathrin: A unique protein associated with intracellular transfer of membrane by coated vesicles. Proc Natl Acad Sci USA 73:1255–1259

Pelletier G (1973) Secretion and uptake of peroxidase by rat adenohypophyseal cells. J Ultrastruct Res 43:445–459

Pelttari A, Helminen HJ (1983) Thickness of membranes in secretory epithelial cells of the rat ventral prostate. Biol Cell 47:343–350

Perez M, Hirschberg CB (1985) Translocation of UDP-N-acetylglucosamine into vesicles derived from rat liver rough endoplasmic reticulum and Golgi apparatus. J Biol Chem 260:4671–4678

Pictet RL, Clark WR, Williams RH, Rutter WJ (1972) An ultrastructural analysis of the developing embryonic pancreas. Dev Biol 29:436-467

Pohlmann R, Waheed A, Hasilik A, von Figura K (1982) Synthesis of phosphorylated recognition marker in lysosomal enzymes is located in the cis part of Golgi apparatus. J Biol Chem 257:5323–5325

Posthuma G, Slot JW, Geuze HJ (1984) Immunocytochemical assays of amylase and chymotrypsinogen in rat pancreas secretory granules. J Histochem Cytochem 32:1028–1034

Quaroni A, Kirsch K, Weiser MN (1979) Synthesis of membrane glycoproteins in rat small-intestinal villus cells. Effect of colchicine on the redistribution of L-[1,5,6-3H]-fucose-labelled membrane glycoproteins among Golgi, lateral basal and microvillus membranes. Biochem J 182:213–221

Quatacker JR (1979) Different aspects of membrane differentiation at the inner side (GERL) of the Golgi apparatus in rabbit luteal cells. Histochem J 11:399–416

Quinn P, Griffiths G, Warren G (1983) Dissection of the Golgi complex. II. Density separation of specific Golgi functions in virally infected cells treated with monensin. J Cell Biol 96:851–856

Rambourg A, Clermont Y (1986) Tridimensional structure of the Golgi apparatus in type A ganglion cells of the rat. Am J Anat 176:393–409

Rambourg A, Hernandez W, Leblond CP (1969) Detection of complex carbohydrates in the Golgi apparatus of rat cells. J Cell Biol 40:395–414

Rambourg A, Clermont Y, Hermo L (1979) Three-dimensional architecture of the Golgi apparatus in Sertoli cells of the rat. Am J Anat 154:455–476

Rambourg A, Clermont Y, Hermo L (1981) Three-dimensional structure of the Golgi apparatus. In: Hand A, Oliver C (eds) Methods in cell biology, vol 23, part 2. Academic, New York, pp 155–166

Rambourg A, Secretain D, Clermont Y (1984) Tridimensional architecture of the Golgi apparatus in the atrial muscle cell of the rat. Am J Anat 170:163–179

Ratcliffe A, Fryer PR, Hardingham TE (1985) Proteoglycan biosynthesis in chondrocytes: Protein A-gold localization of proteoglycan protein core and chondroitin sulfate within Golgi subcompartments. J Cell Biol 101:2355–2365

Ravazzola M, Perrelet A, Roth J, Orci L (1981) Insulin immunoreactive sites demonstrated in the Golgi apparatus of pancreatic B cells. Proc Natl Acad Sci USA 78:5661–5664

Reaven EP, Reaven GM (1977) Distribution and content of microtubules in relation to the transport of lipid. An ultrastructural quantitative study of the absorptive cell of the small intestine. J Cell Biol 75:559–572

Redman CM, Banerjee D, Howell K, Palade GE (1975) Colchicine inhibition of plasma protein release from rat hepatocytes. J Cell Biol 66:42–59

Reggio H, Bainton D, Harms E, Coudrier E, Louvard D (1984) Antibodies against lysosomal membranes reveal a 100,000-mol-wt protein that cross-reacts with purified H^+,K^+-ATPase from gastric mucosa. J Cell Biol 99:1511–1526

Regoeczi E, Chindemi PA, Debanne MT, Charlwood PA (1982) Partial resialylation of human asialotransferrin type 3 in the rat. Proc Natl Acad Sci USA 79:2226–2230

Rindler MJ (1986) Biogenesis of plasma membranes in polarized epithelial cells. Biochem Soc Trans 14:159–161

Rindler MJ, Ivanov IE, Plesken H, Rodriguez-Boulan E, Sabatini DD (1984a) Viral glycoproteins destined for apical or basolateral plasma membrane domains traverse the same Golgi apparatus during their intracellular transport in doubly infected Madin-Darby canine kidney cells. J Cell Biol 98:1304–1319

Rindler MJ, Ivanov IE, Sabatini DD (1984b) Microtubule-acting drugs interfere with the delivery of influenza hemagglutinin (HA) to the apical plasma membrane of MDCK cells. J Cell Biol 99:6a (abstract)

Rindler MJ, Ivanov IE, Plesken H, Sabatini DD (1985) Polarized delivery of viral glycoproteins to the apical and basolateral plasma membranes of Madin-Darby canine kidney cells infected with temperature-sensitive viruses. J Cell Biol 100:136–151

Robbins AR, Oliver C, Bateman JL, Krag SS, Galloway JJ, Mellman I (1984) A single mutation in Chinese hamster ovary cells impairs both Golgi and endosomal functions. J Cell Biol 99:1296–1308

Robinson DG, Kristen U (1982) Membrane flow via the Golgi apparatus of higher plant cells. Int Rev Cytol 77:89–127

Robinson MS, Pearse BMF (1986) Immunofluorescent localization of 100K coated vesicle proteins. J Cell Biol 102:48–54

Rodriguez JL, Gelpi C, Thomson TM, Real FJ, Fernandez J (1982) Anti-Golgi complex autoantibodies in a patient with Sjögren's syndrome and lymphoma. Clin Exp Immunol 49:579–586

Rogalski AA, Singer SJ (1984) Associations of elements of the Golgi apparatus with microtubules. J Cell Biol 99:1092–1100

Rogalski AA, Bergmann JE, Singer SJ (1984) Effect of microtubule assembly on the intracellular processing and surface expression of an integral protein of the plasma membrane. J Cell Biol 99:1101–1109

Romagnoli P (1984) The Golgi apparatus and lysosomes of rat pancreatic acinar cells following refeeding. Histochem J 16:855–868

Rose JK, Bergmann JE (1983) Altered cytoplasmic domains affect intracellular transport of the vesicular stomatitis virus glycoprotein. Cell 34:513–524

Roth J (1983) Application of lectin-gold complexes for electronmicroscopic localization of glycoconjugates on thin sections. J Histochem Cytochem 31:987–999

Roth J (1984) Cytochemical localization of terminal N-acetyl-D-galactosamine residues in cellular compartments of intestinal goblet cells: Implications for the topology of O-glycosylation. J Cell Biol 98:399–406

Roth J, Berger E (1982) Immunocytochemical localization of galactosyltransferase in Hela cells: Codistribution with thiamine pyrophosphatase in trans-Golgi cisternae. J Cell Biol 93:223–229

Roth J, Lucocq JM, Charest PM (1984) Light and electron microscopic demonstration of sialic acid residues with the lectin from Limax flavus: A cytochemical affinity technique with the use of fetuin-gold complexes. J Histochem Cytochem 32:1167–1176

Roth J, Taatjes DJ, Lucocq JM, Weinstein J, Paulson JC (1985) Demonstration of an extensive trans-tubular network continuous with the Golgi apparatus stack that may function in glycosylation. Cell 43:287–295

Roth J, Taatjes DJ, Weinstein J, Paulson JC, Greenwell P, Watkins WM (1986) Differential subcompartmentation of terminal glycosylation in the Golgi apparatus of intestinal absorptive and goblet cells. J Biol Chem 26:14307–14312

Rothman JE (1981) The Golgi apparatus: Two organelles in tandem. Science 213:1212–1219

Rothman JE (1985) The compartmental organization of the Golgi apparatus. Sci Am 253:84–95

Rothman JE, Fine RE (1980) Coated vesicles transport newly synthesized membrane glycoproteins from endoplasmic reticulum to plasma membrane in two successive stages. Proc Natl Acad Sci USA 77:780–784

Rothman JE, Urbani LJ, Brands R (1984a) Transport of protein between cytoplasmic membranes of fused cells: Correspondence to processes reconstituted in a cell free system. J Cell Biol 99:248–259

Rothman JE, Miller RL, Urbani LJ (1984b) Intercompartmental transport in the Golgi complex is a dissociative process: Facile transfer of membrane protein between two Golgi populations. J Cell Biol 99:260–271

Sabesin SM, Frase S (1977) Electron microscopic studies of the assembly, intracellular transport, and secretion of chylomicrons by rat intestine. J Lipid Res 18:496–511

Sage J, Jersild R (1971) Comparative distribution of carbohydrates and lipid droplets in the Golgi apparatus of intestinal absorptive cells. J Cell Biol 51:333–338

Sandoval IV, Bonifacino JS, Klausner RD, Henkart M, Wehland J (1984) Role of microtubules in the organization and localization of the Golgi apparatus. J Cell Biol 99:113s–118s
Saraste J, Kuismanen E (1984) Pre- and post-Golgi vacuoles operate in the transport of Semliki forest virus membrane glycoproteins to the cell surface. Cell 38:535–549
Saraste J, Palade GE, Farquhar MG (1986) Temperature-sensitive steps in the transport of secretory proteins through the Golgi complex in exocrine pancreatic cells. Proc Natl Acad Sci USA 83:6425–6429
Sasaki T (1983) Ultrastructure and cytochemistry of the Golgi apparatus and related organelles of the secretory ameloblasts of the rat incisor. Arch Oral Biol 28:895–905
Sasaki T, Motegi N, Higashi S (1984) Morphological analysis of the Golgi apparatus in rat amelogenesis as revealed by the Ur-Pb-Cu block staining method and freeze-fracture replication. J Electr Microsc 33:19–33
Sato A, Spicer SS (1982a) Ultrastructural visualization of galactosyl residues in various alimentary tract epithelial cells with the peanut lectin-horseradish peroxidase procedure. Histochemistry 73:607–624
Sato A, Spicer SS (1982b) Ultrastructural visualization of galactose in the glycoprotein of gastric surface cells with a peanut lectin conjugate. Histochem J 14:125–138
Sawano F, Fujita H (1981) Cytochemical studies on the internal polarity of the Golgi apparatus and the relationship between this organelle and GERL. Histochemistry 71:335–348
Schachter H, Roseman S (1980) Mammalian glycosyltransferases. Their role in the synthesis and function of complex carbohydrates and glycolipids. In: Lennarz WJ (ed) The biochemistry of glycoproteins and proteoglycans. Plenum, New York, pp 85–160
Scheele G, Tartakoff A (1985) Exit of nonglycosylated secretory proteins from the rough endoplasmic reticulum is asynchronous in the exocrine pancreas. J Biol Chem 260:926–931
Schlossman DM, Schmid SL, Braell WA, Rothman JE (1984) An enzyme that removes clathrin coats: Purification of an uncoupling ATPase. J Cell Biol 99:723–733
Schmid SL, Braell WA, Schlossman DM, Rothman JE (1984) A role of clathrin light chains in the recognition of clathrin cages by "uncoating ATPase". Nature 311:228–231
Schroeter D, Kiesewetter VL, Reggio H (1984) Identification of intracellular membrane systems in mitotic cells by immunocytochemistry. Eur J Cell Biol [Suppl] 5:34 (abstract)
Schroeter D, Eheman V, Paweletz N (1985) Cellular compartments in mitotic cells: Ultrahistochemical identification of Golgi elements in PtK-1 cells. Biol Cell 53:155–164
Schwartz AL, Strous GJAM, Slot JW, Geuze HJ (1985) Immunoelectron microscopic localization of acidic intracellular compartments in hepatoma cells. EMBO J 4:899–904
Schwarz H, Thilo L (1983) Membrane traffic in *Dictyostelium discoideum:* Plasma membrane glycoconjugates internalized and recycled during fluid phase pinocytosis enter the Golgi complex. Eur J Cell Biol 31:212–219
Schwarz JK, Capasso JM, Hirschberg CB (1984) Translocation of adenosine 3′-phosphate 5′-phosphosulfate into rat liver Golgi vesicles. J Biol Chem 259:3554–3559
Scott CA, Flickinger CJ (1983) Secretory process in Brunner's glands during recovery from stimulation with a single dose of pilocarpine. Anat Rec 206:267–282
Sesso A, Nicolosi C, Catena RS, Correa H (1983) Freeze-fracture characterization of the outermost Golgi cisterna (OGC) in rat pancreatic acinar cells. Biol Cell 48:175–184
Sharon N, Lis H (1972) Lectins: Cell agglutinating and sugar-specific proteins. Science 177:949–959
Sleight RG, Pagano RE (1984) Transport of a fluorescent phosphatidylcholin analog from the plasma membrane to the Golgi apparatus. J Cell Biol 99:742–751
Slot JW, Geuze JJ (1979) A morphometrical study of the exocrine pancreatic cell in fasted and fed frogs. J Cell Biol 80:692–707
Slot JW, Geuze HJ (1983) Immunoelectron microscopic exploration of the Golgi complex. J Histochem Cytochem 31:1049–1056
Sly WS, Fischer HD (1982) The phosphomannosyl recognition system for intracellular and intercellular transport of lysosomal enzymes. J Cell Biochem 18:67–85
Smith CE (1980) Ultrastructural localization of nicotinamide adenine dinucleotide phosphatase (NADPase) activity to the intermediate saccules of the Golgi apparatus in rat incisor ameloblasts. J Histochem Cytochem 28:16–26
Smith CE (1981) Ultrastructural localization of coenzyme A phosphatase (CoAPase) activity to the GERL system in secretory ameloblasts of the rat incisor. J Histochem Cytochem 29:1243–1254

Smith CE, Paiement J, Bergeron JJM (1986) Subcellular distribution of acid NADPase activity within the parenchymal cells of rat liver. J Histochem Cytochem 34:649–658

Smith RE, Farquhar MG (1970) Modulation of nucleoside diphosphatase activity of mammotrophic cells of the rat adenohypophysis during secretion. J Histochem Cytochem 18:237–250

Smith ZDJ, D'Eugenio-Gumkowski F, Yanagisawa K, Jamieson JD (1984) Endogenous and monoclonal antibodies to the rat pancreatic acinar cell Golgi complex. J Cell Biol 98:2035–2046

Snider MD, Rogers OC (1985) Intracellular movement of cell surface receptors after endocytosis: Resialylation of asialo-transferrin receptor in human erythroleukemia cells. J Cell Biol 100:826–834

Snider MD, Rogers OC (1986) Membrane traffic in animal cells: Cellular glycoproteins return to the site of Golgi mannosidase I. J Cell Biol 103:265–275

Spater HW, Novikoff AB, Spater SH, Quintana N (1978) Pyridoxal phosphatase: cytochemical localization in GERL and other organelles of rat neurons. J Histochem Cytochem 26:809–821

Spicer SS, Schulte BA (1982) Ultrastructural methods for localizing complex carbohydrates. Human Pathol 13:343–354

Staehelin LA, Kiermayer O (1970) Membrane differentiation in the Golgi complex of *Micrasterias denticulata Bréb.* visualized by freeze-etching. J Cell Sci 7:787–792

Stanka P, Rathjen P, Sahlmann B (1981) Evidence of membrane transformation during melanogenesis. Electronmicroscopic study on the retinal pigment epithelium of chick embryos. Cell Tiss Res 214:343–353

Steinman RM, Mellman IS, Muller WA, Cohn ZA (1983) Endocytosis and the recycling of plasma membrane. J Cell Biol 96:1–27

Stenseth K, Thyberg J (1986) Pathways of endocytosis in cultured macrophages. Electron microscopic autoradiographic tracing of iodinated plasma membrane proteins. Eur J Cell Biol 40:37–43

Sterle M, Pipan N (1985) The analysis of GA in stomach mucoid cells of mouse embryo during development. Hungarian-Austrian Joint Conference on Electron Microscopy, April 25–27, Batatonaliga, Hungary, p 94 (abstract)

Stieber A, Gonatas JO, Gonatas NK (1984) Differences between the endocytosis of horseradish peroxidase and its conjugate with wheat germ agglutinin by cultured fibroblasts. J Cell Physiol 119:71–76

Storrie B, Pool RR Jr, Sachdeva M, Maurey KM, Oliver C (1984) Evidence for both prelysosomal and lysosomal intermediates in endocytic pathways. J Cell Biol 98:108–115

Stow JL, Kjellén L, Unger E, Höök M, Farquhar MG (1985) Heparan sulfate proteoglycans are concentrated on the sinusoidal plasmalemmal domain and in intracellular organelles of hepatocytes. J Cell Biol 100:975–980

Strous GJAM, Berger EG (1982) Biosynthesis, intracellular transport, and release of the Golgi enzyme galactosyltransferase (lactose synthetase A protein) in HeLa cells. J Biol Chem 257:7623–7628

Strous GJ, Kerkhof P van, Willemsen R, Geuze HJ, Berger EG (1983a) Transport and topology of galactosyltransferase in endomembranes of HeLa cells. J Cell Biol 97:723–727

Strous GJAM, Willemsen R, Kerkhof P van, Slot JW, Geuze HJ, Lodish HF (1983b) Vesicular stomatitis virus glycoprotein, albumin, and transferrin are transported to the cell surface via the same Golgi vesicles. J Cell Biol 97:1915–1922

Strous GJ, DuMaine A, Zijderhand-Bleekemolen JE, Slot JW, Schwartz AL (1985a) Effect of lysosomotropic amines on the secretory pathway and on the recycling of the asialoglycoprotein receptor in human hepatoma cells. J Cell Biol 101:531–539

Strous GJ, Kerkhof P van, Willemsen R, Slot JW, Geuze HJ (1985b) Effect of monensin on the metabolism, localization, and biosynthesis of N- and O-linked oligosaccharides of galactosyltransferase. Eur J Cell Biol 36:256–262

Sturgess J, Moscarello M, Schachter H (1978) The structure and biosynthesis of membrane glycoproteins. Curr Top Membrane Transport 11:15–105

Susi FR, Leblond CP, Clermont Y (1971) Changes in the Golgi apparatus during spermiogenesis in the rat. Am J Anat 130:251–268

Tanaka K, Mitsushima A, Fukudome H, Kashima Y (1986) Three-dimensional architecture of the Golgi complex observed by high resolution scanning electron microscopy. J Submicrosc Cytol 18:1–9

Tang XM, Lalli MF, Clermont Y (1982) A cytochemical study of the Golgi apparatus of the spermatid during spermiogenesis in the rat. Am J Anat 163:283–294

Tartakoff AM (1980) The Golgi complex. Crossroads for vesicular traffic. In: Richter GW, Epstein AM (eds) International review of experimental pathology, vol 22. Academic, New York, pp 227–251

Tartakoff AM (1983a) The confined function model of the Golgi complex: Center for ordered processing of biosynthetic products of the rough endoplasmic reticulum. Int Rev Cytol 85:221–252

Tartakoff AM (1983b) Perturbation of vesicular traffic with the carboxylic ionophore monensin. Cell 32:1026–1028

Tartakoff AM (1986) Temperature and energy dependence of secretory protein transport in the exocrine pancreas. EMBO J 5:1477–1482

Tartakoff AM, Vassalli P (1977) Plasma cell immunoglobulin secretion. Arrest is accompanied by alterations of the Golgi complex. J Exp Med 146:1332–1345

Tartakoff AM, Vassalli P (1983) Lectin-binding sites as markers of Golgi subcompartments: Proximal-to-distal maturation of oligosaccharides. J Cell Biol 97:1243–1248

Tartakoff A, Vassalli P, Detraz M (1978) Comparative studies of intracellular transport of secretory proteins. J Cell Biol 79:694–707

Tassin AM, Paintrand M, Berger EG, Bornens M (1985) The Golgi apparatus remains associated with microtubule organizing centers during myogenesis. J Cell Biol 101:630–638

Thyberg J (1980) Internalization of cationized ferritin into the Golgi complex of cultured mouse peritoneal macrophages. Effects of colchicine and cytochalasin B. Eur J Cell Biol 23:95–103

Thyberg J, Stenseth K (1981) Endocytosis of native and cationized horseradish peroxidase by cultured mouse peritoneal macrophages. Variations in cell surface binding and intracellular traffic and effects of colchicine. Eur J Cell Biol 25:308–318

Thyberg J, Moskalewski S (1985) Microtubules and the organization of the Golgi complex. Exp Cell Res 159:1–16

Thyberg J, Piasek A, Moskalewski S (1980) Effect of colchicine on the Golgi complex and GERL of cultured rat peritoneal macrophages and epiphyseal chondrocytes. J Cell Sci 45:42–58

Thyberg J, Blomgren K, Hellgren D, Hedin U (1982) Lysosomophagy in cultured macrophages treated with the antimicrotubular drug nocodazole. Eur J Cell Biol 27:279–288

Tokumitsu S-I, Fishman WH (1983) Alkaline phosphatase biosynthesis in the endoplasmic reticulum and its transport through the Golgi apparatus to the plasma membrane. Cytochemical evidence. J Histochem Cytochem 31:647–655

Tougard C, Louvard D, Picart R, Tixier-Vidal A (1983) The rough endoplasmic reticulum and the Golgi apparatus visualized using specific antibodies in normal and tumoral prolactin cells in culture. J Cell Biol 96:1197–1207

Tougard C, Louvard D, Picart R, Tixier-Vidal A (1985) Antibodies against a lysosomal membrane antigen recognize a prelysosomal compartment involved in the endocytic pathway in cultured prolactin cells. J Cell Biol 100:786–793

Treilhou-Lahille F (1982) The secretory process of the adult mouse thyroid "C" cells and the establishment of the secretory period during fetal life. Biol Cell 43:103–120

Uchiyama Y (1982) The membrane association among the rough- and smooth-surfaced endoplasmic reticulum and Golgi apparatus in rat hepatocytes at certain circadian stages. Tohoku J Exp Med 136:299–302

Uchiyama Y (1983) A histochemical study of variations in the localization of 5′-nucleotidase activity in the acinar cell of the rat exocrine pancreas over the twenty-four hour period. Cell Tiss Res 230:411–420

Uchiyama Y, Saito K (1982) A morphometric study of 24-hour variations in subcellular structures of the rat pancreatic acinar cells. Cell Tiss Res 226:609–620

van den Bosch R, Geuze HJ, Strous GJ (1986) Presence of asialoglycoprotein receptors in the Golgi complex in the absence of protein synthesis. Exp Cell Res 162:231–242

van Deurs B, Tonnessen TI, Petersen OW, Sandvig K, Olsnes S (1986) Routing of internalized ricin and ricin conjugates in the Golgi complex. J Cell Biol 102:37–47

van Dongen JM, Barneveld RA, Geuze HJ, Galjaard H (1984) Immunocytochemistry of lysosomal hydrolases and their precursor forms in normal and mutant human cells. Histochem J 16:941–954

Vassy J, Rissel M, Kraemer M, Foucrier J, Guillouzo A (1984) Ultrastructural indirect immunolocalization of transferrin in cultured rat hepatocytes permeabilized with saponin. J Histochem Cytochem 32:538–540

Völkl A (1980) Membranumsatz beim Export von Proteinen. Die Rolle des Golgi Apparates (Referat). Verh Anat Ges 74:97–111

Völkl A, Bieger W, Kern HF (1976) Studies on secretory glycoproteins in the rat exocrine pancreas. I. Fine structure of the Golgi complex and release of fucose-labeled proteins after in vivo stimulation with caerulein. Cell Tissue Res 175:227–243

Volkmann D (1983) A freeze-fracture study on the differentiation of Golgi and plasma membranes in plant cells. Eur J Cell Biol 30:258–265

Wagner DD, Mayadas T, Marder VJ (1986) Initial glycosylation and acidic pH in the Golgi apparatus are required for multimerization of the von Willebrand factor. J Cell Biol 102:1320–1324

Wakayama M, Matsuzaki H, Nakai Y (1983/1984) Golgi apparatus: Distinct structure of acid phosphatase localization in regenerating human skeletal muscle fiber. Dev Neurosci 6:152–160

Wang J-J, Moller PC, Chang JP (1983) Localization of lectin receptors in GERL of Ehrlich ascites tumor cells. Biol Cell 47:285–290

Warren G (1985) Membrane traffic and organelle division. Trends Biochem Sci 10:439–443

Wattenberg BW, Balch WE, Rothman JE (1986) A novel prefusion complex formed during protein transport between Golgi cisternae in a cell-free system. J Biol Chem 261:2202–2207

Weakley BS, Webb P, James JL (1981) Cytochemistry of the Golgi apparatus in developing ovarian germ cells of the Syrian hamster. Cell Tissue Res 220:349–372

Wehland J, Sandoval IV (1983) Cells injected with guanosine 5′-(α-β-methylene) triphosphate, an α,β-non-hydrolyzable analog of GTP, show anormalous patterns of tubulin polymerization affecting cell translocation, intracellular movement, and the organization of Golgi elements. Proc Natl Acad Sci USA 80:1938–1941

Wehland J, Willingham MC (1983) A rat monoclonal antibody reacting specifically with the tyrosylated form of α-tubulin. II. Effects on cell movement, organization of microtubules, and intermediate filaments, and arrangement of Golgi elements. J Cell Biol 97:1476–1490

Wehland J, Henkart M, Klausner R, Sandoval IV (1983) Role of microtubules in the distribution of the Golgi apparatus: Effect of taxol and microinjected anti-α-tubulin antibodies. Proc Natl Acad Sci USA 80:4286–4290

Whaley WG (1975) The Golgi apparatus. Springer, New York (Cell biology monographs vol 2)

Whaley WG, Dauwalder M (1979) The Golgi apparatus, the plasma membrane, and functional integration. Int Rev Cytol 58:199–245

Widnell CC, Schneider Y-J, Pierre B, Baudhuin P, Trouet A (1982) Evidence for a continual exchange of 5′-nucleotidase between the cell surface and cytoplasmic membranes in cultured rat fibroblasts. Cell 28:61–70

Williams JA (1981) Effects of antimitotic agents on ultrastructure and intracellular transport of protein in pancreatic acini. In: Hand A, Oliver C (eds) Methods in cell biology, vol 23. Academic, New York, pp 247–258

Williams LM, Lafontaine JC (1985) An electron microscope thick section study of endomembrane organization in the *Myxamoebae* of *Physorum polycephalum*. Protoplasma 124:42–51

Willingham MC, Pastan I (1980) The receptosome: An intermediate organelle of receptor-mediated endocytosis in cultured fibroblasts. Cell 21:67–77

Willingham MC, Pastan I (1982) Transit of epidermal growth factor through coated pits of the Golgi system. J Cell Biol 94:207–212

Willingham MC, Pastan I (1984) Endocytosis and exocytosis: Current concepts of vesicle traffic in animal cells. Int Rev Cytol 92:51–92

Willingham MC, Pastan I (1985a) Ultrastructural immunocytochemical localization of the transferrin receptor using a monoclonal antibody in human KB cells. J Histochem Cytochem 33:59–64

Willingham MC, Pastan I (1985b) Receptosomes, endosomes, CURL: Different terms for the same organelle systems. Trends Biochem Sci 10:190–191

Willingham MC, Maxfield FR, Pastan I (1980) Receptor-mediated endocytosis of alpha-2-macroglobulin in cultured fibroblasts. J Histochem Cytochem 28:818–823

Willingham MC, Keen JH, Pastan IH (1981a) Ultrastructural immunocytochemical localization of clathrin in cultured fibroblasts. Exp Cell Res 132:329–338

Willingham MC, Pastan IH, Sahagian GG, Jourdian GW, Neufeld EF (1981b) Morphologic study of the internalization of a lysosomal enzyme by the mannose-6-phosphate receptor in cultured Chinese hamster ovary cells. Proc Natl Acad Sci USA 78:6967–6971

Willingham MC, Haigler HT, Fitzgerald DJP, Gallo MC, Rutherford AV, Pastan IH (1983a) The morphologic pathway of binding and internalization of epidermal growth factor in cultured cells. Exp Cell Res 146:163–175

Willingham MC, Pastan IH, Sahagian GG (1983b) Ultrastructural immunocytochemical localization of the phosphomannosyl receptor in Chinese hamster ovary (CHO) cells. J Histochem Cytochem 31:1–11

Willingham MC, Hanover JA, Dickson RB, Pastan I (1984) Morphologic characterization of the pathways of transferrin endocytosis and recycling in human KB cells. Proc Natl Acad Sci USA 81:175–179

Wolkoff AW, Klausner RD, Ashwell G, Harford J (1984) Intracellular segregation of asialoglycoproteins and their receptor: A prelysosomal event subsequent to dissociation of the ligand-receptor complex. J Cell Biol 98:375–381

Woods JW, Doriaux M, Farquhar MG (1986) Transferrin receptors recycle to cis and middle as well as trans Golgi cisternae in Ig-secreting myeloma cells. J Cell Biol 103:277–286

Wright GM, Leblond CP (1981) Immunohistochemical localization of procollagens. III. Type I procollagen antigenicity in osteoblasts and prebone (osteoid). J Histochem Cytochem 29:791–804

Yamamoto A, Masaki R, Tashiro Y (1985) Is cytochrome P-450 transported from the endoplasmic reticulum to the Golgi apparatus in rat hepatocytes? J Cell Biol 101:1733–1740

Yamashiro DJ, Tycko B, Fluss SR, Maxfield FR (1984) Segregation of transferrin to a mildly acidic (pH 6.5) para-Golgi compartment in the recycling pathway. Cell 37:789–800

Yamazaki M, Hayaishi O (1968) Allosteric properties of nucleoside diphosphatase and its identity with TPPase. J Biol Chem 243:2934–2942

Yokota S, Fahimi HD (1981) Immunocytochemical localization of albumin in the secretory apparatus of rat liver parenchymal cells. Proc Natl Acad Sci USA 78:4970–4974

Young RW (1973) The role of the Golgi apparatus in the sulfate metabolism. J Cell Biol 57:175–189

Yusuf HKM, Pohlentz G, Schwarzmann G, Sandhoff K (1983) Ganglioside biosynthesis in Golgi apparatus of rat liver. Stimulation by phosphatidylglycerol and inhibition by tunicamycin. Eur J Biochem 134:47–54

Yusuf HKM, Pohlentz G, Sandhoff K (1984) Ganglioside biosynthesis in Golgi apparatus: New perspectives on its mechanism. J Neurosci Res 12:161–178

Zang F, Schneider DL (1983) The bioenergetics of Golgi apparatus function: Evidence for an ATP-dependent proton pump. Biochem Biophys Res Commun 114:620–625

Zeligs JD, Wollman SH (1979) Mitosis in rat thyroid epithelial cells in vivo. I. Ultrastructural changes in cytoplasmic organelles during the mitotic cycle. J Ultrastruct Res 66:53–77

Addendum

Since completion of the manuscript, several studies being particularly interesting with respect to the Golgi apparatus organization have been published:

Allan VJ, Kreis TE (1986) A microtubule-binding protein associated with membranes of the Golgi apparatus. J Cell Biol 103:2229–2239

Balch WE, Wagner KR, Keller DS (1987) Reconstitution of transport of vesicular stomatitis virus G protein from the endoplasmic reticulum to the Golgi complex using a cell-free system. J Cell Biol 104:749–760

Balin BJ, Broadwell RD (1987) Lectin-labeled membrane is transferred to the Golgi complex in mouse pituitary cells in vivo. J Histochem Cytochem 35:489–498

Barriocanal JG, Bonifacino JS, Yuan L, Sandoval IV (1986) Biosynthesis, glycosylation, movement through the Golgi system, and transport to lysosomes by an N-linked carbohydrate-independent mechanism of three lysosomal integral membrane proteins. J Biol Chem 261:16755–16763

Bennett G, Haddad A (1986) Synthesis and migration of ^{3}H-fucose-labeled glycoproteins in the ciliary epithelium of the eye: Effects of microtubule-disrupting drugs. Amer J Anat 177:441–455

Böck P (1986) Histochemistry of the carotid body: Fine-structural localization of acid phosphatase activities. Neurohistochemistry: Modern Methods and Applications: 617–628

Bourguignon LYW, Balazovich K, Suchard SJ, Hindsgaul O, Pierce M (1986) Endocytosis of mannose-6-phosphate binding sites by mouse T-lymphoma cells. J Cell Physiol 127:146–161

Burgos MH, Gutiérrez LS (1986) The Golgi complex of the early spermatid in guinea pig. Anat Rec 216:139–145

Faye L, Sturm A, Bollini R, Vitale A, Chrispeels MJ (1986) The position of the oligosaccharide side-chains of phytohemagglutinin and their accessibility to glycosidases determines their subsequent processing in the Golgi. Eur J Biochem 158:655–661

Fishman JB, Cook JS (1986) The sequential transfer of internalized, cell surface sialoglycoconjugates through the lysosomes and Golgi complex in Hela cells. J Biol Chem 261:11896–11905

Gahmberg N, Pettersson RF, Kääriäinen L (1986) Efficient transport of Semliki forest virus glycoproteins through a Golgi complex morphologically altered by Uukuniemi virus glycoproteins. EMBO J 5:3111–3118

Gonatas NK, Gonatas JO, Stieber A, Ternynck T, Avrameas S (1987) Detection of plasma cell immunoglobulins in tissue sections optimally fixed for ultrastructural immunocytochemistry. J Histochem Cytochem 35:189–196

Griffiths G, Simons K (1986) The trans Golgi network: Sorting at the exit site of the Golgi complex. Science 234:438–443

Hedman K, Goldenthal KL, Rutherford AV, Pastan I, Willingham MC (1987) Comparison of the intracellular pathway of transferrin recycling and vesicular stomatitis virus membrane glycoprotein exocytosis by ultrastructural double-label cytochemistry. J Histochem Cytochem 35:233–243

Hoflack B, Fujimoto K, Kornfeld S (1987) The interaction of phosphorylated oligosaccharides and lysosomal enzymes with bovine liver cation-dependent mannose-6-phosphate receptor. J Biol Chem 262:123–129

Kanwar YS, Rosenzweig LJ, Jakubowski ML (1986) Xylosylated-proteoglycan-induced Golgi alterations. Proc Natl Acad Sci USA 83:6499–6503

Lis H, Sharon N (1986) Lectins as molecules and as tools. Ann Rev Biochem 55:35–67

Lucocq JM, Berger EG, Roth J (1987) Detection of terminal N-linked N-acetylglucosamine residues in the Golgi apparatus using galactosyltransferase and endoglucosaminidase F/ peptide N-glycosidase F: Adaption of a biochemical approach to electron microscopy. J Histochem Cytochem 35:67–74

Malchiodi F, Rambourg A, Clermont Y, Caroff A (1986) Ultrastructural localization of Concanavalin A-binding sites in the Golgi apparatus of various types of neurons in rat dorsal root ganglia: Functional implications. Amer J Anat 177:81–95

Mellman I, Fuchs R, Helenius A (1986) Acidification of the endocytic and exocytic pathways. Ann Rev Biochem 55:663–700

Minnifield N, Creek KE, Navas P, Morré DJ (1986) Involvement of cis and trans Golgi apparatus elements in the intracellular sorting and targeting of acid hydrolases to lysosomes. Eur J Cell Biol 42:92–100

Murata F, Suganuma T, Tsuyama S, Ishida KS, Funasako S (1986) Glycoconjugate cytochemistry of the rat small intestine using Helix pomatia agglutinin and colloidal gold conjugates. J Electron Microsc 35:29–37

Nuwayhid N, Glaser JH, Johnson JC, Conrad HE, Hauser SC, Hirschberg CB (1986) Xylosylation and glucuronosylation reactions in rat liver Golgi apparatus and endoplasmic reticulum. J Biol Chem 261:12936–12941

Orci L (1986) The insulin cell: Its cellular environment and how it processes (pro)-insulin. Diabetes/Metabolism Rev Vol 2:71–106

Orci I, Ravazzola M, Amherdt M, Brown D, Perrelet A (1986) Transport of horseradish peroxidase from the cell surface to the Golgi in insulin-secreting cells: Preferential labelling of cisternae located in an intermediate position in the stack. EMBO J 5:2097–2101

Orci L, Ravazzola M, Amherdt M, Madsen O, Perrelet A, Vassalli J-D, Anderson RGW (1986) Conversion of proinsulin to insulin occurs coordinately with acidification of maturing secretory vesicles. J Cell Biol 103:2273–2281

Paiement J (1986) Morphology of endoplasmic reticulum and Golgi elements following microinjection of rat liver microsomes into Xenopus laevis oocyte cytoplasm. Exp Cell Res 166:510–518

Parenti C, Willemsen R, Hoogeveen AT, Verleun-Mooyman M, van Dongen JM, Galjaard H (1987) Immunocytochemical localization of lysosomal acid phosphatase in normal and "I-cell" fibroblasts. Eur J Cell Biol 43:121–127

Perez M, Hirschberg CB (1986) Transport of sugar nucleotides and adenosine 3′-phosphate 5′-phosphosulfate into vesicles derived from the Golgi apparatus Biochim Biophys Acta 864:213–222

Roth J (1987) Subcellular organization of glycosylation in mammalian cells. Biochim Biophys Acta, Rev on Biomembranes, in press

Renan-Piqueras J, Miragall F, Guerri C, Baguena-Cervellera R (1987) Prenatal exposure to alcohol alters the Golgi apparatus of newborn rat hepatocytes: A cytochemical study. J Histochem Cytochem 35:221–228

Strous GJ (1986) Golgi and secreted galactosyltransferase. CRC Critical Rev on Biochemistry 21:119–151

Taatjes DJ, Roth J (1986) The trans-tubular network of the hepatocyte Golgi apparatus is part of the secretory pathway. Eur J Cell Biol 42:344–350

Vila-Porcile E, Picart R, Tixier-Vidal A, Tougard C (1987) Cellular and subcellular distribution of laminin in adult rat anterior pituitary. J Histochem Cytochem 35:287–299

Subject Index*

* Numbers in italics refer to figure legends.

Advances in Anatomy, Embryology and Cell Biology

Editors: F. Beck, W. Hild, W. Kriz, R. Ortmann, J. E. Pauly, T. H. Schiebler

Volume 105
M. Bergmann
Photoperiod and Testicular Function in Phodopus sungorus
1987. 37 figures. VI, 76 pages. ISBN 3-540-17583-0

Volume 104
J. M. de Kock
The Development of the Chondrocranium of Melopsittacus undulatus
1987. 40 figures. VI, 70 pages. ISBN 3-540-17674-8

Volume 103
E. B. Krammer, M. F. Lischka, T. P. Egger. M. Riedl, H. Gruber
The Motoneuronal Organization of the Spinal Accessory Nuclear Complex
1987. 11 figures. IX, 62 pages. ISBN 3-540-17459-1

Volume 102
D. J. Pallot
The Mammalian Carotid Body
1987. 35 figures. VI, 92 pages. ISBN 3-540-17480-X

Volume 101
A. Kress, J. Millian
The Female Genital Tract of the Shrew Crocidura russula
1987. 31 figures. VI, 76 pages. ISBN 3-540-16942-3

Volume 100
J. Altmanm,, S. A. Bayer
The Development of the Rat Hypothalamus
1986. 125 figures. XV, 178 pages. ISBN 3-540-16654-8

Volume 99
W. F. Neiss
Ultracytochemistry of Intracellular Membrane Glycoconjugates
1986. 113 figures. VIII, 92 pages. ISBN 3-540-16726-9

Volume 98
K. Hinrichsen
The Early Development of Morphology and Patterns of the Face in the Human Embryo
1985. 82 figures. VII, 79 pages. ISBN 3-540-15848-0

Springer-Verlag
Berlin Heidelberg New York
London Paris Tokyo

Springer

Advances in Anatomy, Embryology and Cell Biology

Editors: F. Beck, W. Hild, W. Kriz, R. Ortmann, J. E. Pauly, T. H. Schiebler

Volume 97
D. K. Morest, J. A. Winer

The Comparative Anatomy of Neurons: Homologous Neurons in the Medial Geniculate Body of the Opossum and the Cat

1986. 43 figures. XI, 96 pages. ISBN 3-540-15726-3

Volume 96
C. L. Veenman, K.-M. Gottschaldt

The Nucleus Basalis-Neostriatum Complex in the Goose (Anser anser L.)

1986. 41 figures. VIII, 85 pages. ISBN 3-540-15338-1

Volume 95
M. Köhncke

The Chondrocranium of Cryptoprocta ferox

1985. 21 figures. VI, 89 pages. ISBN 3-540-15337-3

Volume 94
B. Demes

Biomechanics of the Primate Skull Base

1985. 29 figures. V, 59 pages. ISBN 3-540-15290-3

Volume 93
M. Rickmann, J., R. Wolff

Prenatal Gliogenesis in the Neopallium of the Rat

1985. 29 figures. VII, 104 pages. ISBN 3-540-13849-8

Volume 92
J. J. Heimans, J. Valk, A. H. M. Lohman

Angiographic Anatomy of the Anterior Inferior Cerebellar Artery

1985. 69 figures. VII, 93 pages. ISBN 3-540-13768-8

Volume 91
A. A. M. Gribnau, L. G. M. Geijsberts

Morphogenesis of the Brain in Staged Rhesus Monkey Embryos

1985. 19 figures. VIII, 69 pages. ISBN 3-540-13709-2

Volume 90
A. J. Verbout

The Development of the Vertebral Column

1985. 71 figures. VII, 122 pages. ISBN 3-540-13730-0

Springer-Verlag
Berlin Heidelberg New York
London Paris Tokyo

Springer